Escaping The Rat Race – Freedom In Paradise

Real-Life Stories About Living, Working, Investing and Retiring in Belize

by

Helga Peham

First published in USA in 2007
by World Audience

Copyright © Helga Peham, 2007

Contact the author at helga.peham@chello.at for comments, questions, film rights, translation rights, TV or other rights.

Copyright notice: All work contained within (including photographs, except front cover crowd image, © Pavel Losevsky) is the sole copyright of its author, 2007, and may not be reproduced without consent.

World Audience (www.worldaudience.org) is a global consortium of artists and writers, producing the literary journal *audience* and *The audience Review*. Our periodicals and books are edited by M. Stefan Strozier and assistant editors. Please submit your stories, poems, paintings, photography, or other artwork, to submissions@worldaudience.org. Inquire about being a reviewer: theatre@worldaudience.org. Thank you.

ISBN 978-1-934209-93-6

10-digit ISBN 1-934209-93-7

Design & Layout: Matthew Ward

New York (NY, USA)

Newcastle (NSW, Australia)

For Gigi

Contents

THANK YOU	i
ABOUT THE AUTHOR	ii
INTRODUCTION: HOW I CAME TO WRITE THIS BOOK	iv
PART I: BUSINESS	1
Chapter 1: AN AGING HIPPIE – PAMELA THE COOKIE LADY	3
Chapter 2: THE BIRD AND HIS WINGS	8
Chapter 3: YOUNG GRINGO LOST AND FOUND HER HEART IN BELIZE	12
Chapter 4: FROM NEWS TO FROZEN CUSTARD	24
Chapter 5: ARCHAEOLOGY AND GOURMET FOOD	32
Chapter 6: HOTEL MANAGEMENT AND BUSINESS RECOMMENDATIONS	35
Chapter 7: WORLD TRAVELER'S BAREFOOT BOOKS	43
Chapter 8: STANDING ON HER OWN FEET	47
Chapter 9: A LAWYER PLAYS MINI GOLF AND SELLS SOFT ICE	56
Chapter 10: A DREAM LIFE	60
Chapter 11: NEWS FROM SAN PEDRO	67
PART II: BELIZEAN FRIENDS	83
Chapter 12: CARING FOR DOGS AND CATS	80
Chapter 13: THE MAYOR IS A LADY	87
Chapter 14: ONE OF THE OLDEST PEOPLE IN SAN PEDRO	93
Chapter 15: "A PARADISE WITH A PAST"	100
Chapter 16: FOUR YOUNG PEOPLE AND A NEWSPAPER	109
Chapter 17: SURROUNDED BY CHILDREN, THE ISLAND ACADEMY	118
Chapter 18: A DILIGENT HIGH SCHOOL STUDENT	123

PART III: REAL ESTATE	131
Chapter 19: TIMESHARE – SHARED TIME	133
Chapter 20: WITH A SMILE ON HER LIPS	140
Chapter 21: FROM HEALTH TO REAL ESTATE SERVICES	149
Chapter 22: FRACTUAL OWNERSHIP MARKETING AND MANAGEMENT	158
Chapter 23: A UKRAINIAN AMERICAN	163
PART IV: ARTISTS	179
Chapter 24: BELIZEAN ARTS	181
Chapter 25: A CARIBBEAN MUSICIAN	186
Chapter 26: AN ARTIST'S LIFE	197
Chapter 27: CORNELIUS MAGNUS HARRELL	201
PART V: SCHOOL AND UNIVERSITY	235
Chapter 28: AESTHETIC PHOTOGRAPHY	237
Chapter 29: SUNSHINE AND MOONLIGHT	243
Chapter 30: CHILDREN AND CHOCOLATE	239
Chapter 31: KIDS, KINESOLOGY, CARBOHYDRATES AND CAYE CAULKER	279
PART VI: RETIREES	299
Chapter 32: A POLICE OFFICER RETIRES	301
Chapter 33: AT HOME IN BELIZE AND THE U.S.	310
Chapter 34: A PARTLY RETIRED LAWYER LIVES HIS DREAM LIFE	318

THANK YOU

My sincere thank you goes to the people I interviewed – with great pleasure – for sharing their stories and wisdom. They made it possible for me to write this book and to learn from them.

A deep thank you to Professor Crandall Hutchkins for allowing me to use so many beautiful photos from his documentary on Ambergris Caye and San Pedro. They are under his sole copyright.

My personal thanks go to Ms. Edna Williams, Mr. Mag. Hans-Peter Gaede, Ms. Mag. Helma Edmond-Malzer and Dr. Alan Morgan who were so kind as to read at various stages a draft of this manuscript. Their valuable comments made it possible for me to improve the quality of the book.

My very special thank you goes to my personal editors Mr. Michael LaRocca and his wife Jan for their careful and professional editing of the several versions of my manuscript.

I am very grateful for the support of the World Audience publishers Mr. Michael Strozier and Mr. Matt Ward, for all the excellence, thought and support they put towards the production and publication of my fourth book.

I want to thank you, dear reader, for reading this book full of sunshine and news for everyone who wants to escape the rat race and dreams of a change, of a journey.

You are very welcome to contact me through my email helga.peham@chello.at with any comments or questions you may have.

ABOUT THE AUTHOR

Translator, teacher, author, world traveler and Belizean resident. Austrian author Dr. Helga Peham brings a unique perspective to Escaping the Rat Race.

Dr. Peham has been working internationally since 1966, when she was a translator (English, French, German) at the Vienna Airport Authorities. She worked for two United Nations organizations as an International Civil Servant in the Training department for more than 20 years. She has translated several books from German into English and from English and French into German in addition to writing several published books of her own. Helga Peham established and managed her own Language Academy, Euro-Languages, and taught English to adults and children.

In January 2005, she moved to Belize, where she worked for one year as the "Director of Scholastic Affairs" in a US start-up medical school, "InterAmerican School of Medical Sciences."

Dr. Peham studied in Suez, Egypt; Vienna, Austria; and Duisburg, Germany before attending Norwood Technical College in London, England. While working at the International Atomic Energy Agency in Vienna, Austria, she studied History and Psychology, completed a PhD in 1981, and also studied English Language and Literature at Vienna University. There she also did further studies in International Politics, Journalism and Communication Sciences for two years. She also pursued postgraduate studies in International

Affairs, International Business, Management, and Business Consulting.

With this extensive educational, writing, and working background, she came to Belize, and it was only natural that she would write about it. She met the people who lived on the island and came to know their stories, and now she will share them with you.

You may find yourself longing to relax amidst white sands, blue oceans, warm tropical breezes, the taste and smell of clean salty air...

INTRODUCTION: HOW I CAME TO WRITE THIS BOOK

WHY I WROTE THIS BOOK

Many people dream of living in a tropical country with sandy beaches and palm trees. I decided to write about people who had realized this dream, so others could learn how they did it. What they did, you can do, if you want.

Belize is a small country on the Yucatan Peninsula, just south of Mexico, by the sea alongside the Barrier Reef, which is the second largest reef in the world. It has the tallest waterfall in Central America and the world's only Jaguar preserve.

When I first came to Belize for 6 weeks in January/February 2003 and again for 45 days in September/October 2004, I was so impressed by the interesting lives of so many people, Belizeans and expatriates, that I, as a biographer, promised myself I would return to interview them and record their stories for others to enjoy. After moving to Belize in mid-January 2005, I was ready to start the project in 2006 and completed it in 2007.

HOW I CAME TO BELIZE

In 2002 I was researching on the Internet when I came across a link leading to Belize. My curiosity was stirred and I Googled "Belize" and related terms for a year before going there to see for myself. Over the Internet, I booked a little cabana in a small resort in Copperbank, in Corozal District. I enjoyed sitting at the lagoon under a palm tree reading and writing and relaxing, in this place I had found by sheer chance.

I grew very interested in Corozal District and the life there because of its proximity to Mexico and the city of Chetumal and because this area, in the north of Belize, has less rain in the rainy season than the southern districts.

A private four-day tour with a teacher from a neighbouring village took me to Cayo and the Mayan pyramids of Xuantunich, the waterfalls, the huge caves and the tropical rain forest in the mountains, and I loved what I saw.

When I returned to Belize for another holiday during the rainy season in 2004, after a few weeks in Copperbank I visited the island of Ambergris Caye with the town of San Pedro. I loved the sandy beaches, the beautiful sea, the wonderful hotels and resorts and the happy people, both Belizeans and expatriates, as well as tourists.

When I received an offer to work as the Director of Scholastic Affairs for a startup US medical school in Corozal Town, Belize, I took it, relocating in January 2005 and working there for one year.

MY RELATIONSHIP WITH THE PEOPLE INTERVIEWED

Whenever I had some time, I went for a few days' holiday to enjoy San Pedro. In February 2006, partly through the help of Professor Floyd Jackson, MD, PhD, who kindly introduced me to his friends, I came into contact with many expatriates, mainly from the US and Canada, and Belizeans living in San Pedro, who were willing to grant me extensive interviews. I wrote the book about their experiences in Belize.

THE COUNTRY OF BELIZE

You may wonder what Belize is like. This English-speaking Caribbean country – a British colony until 1981 and still a member of the British Commonwealth of Nations – is situated on the Yucatan Peninsula, just south of Mexico. It only has about 280,000 inhabitants, with roughly the same number of Belizeans living abroad, mostly in the United States of America. The climate is tropical with a rainy season (June to November) and a dry season (December to May). You will find Mayan ruins, wildlife, tropical fruits, opportunities for diving (especially on the cayes), and kind inhabitants. At school English, the official language, is used. This is also a reason why people love to retire in Belize.

In the northern part of Belize, especially in Corozal District, many people speak Spanish, because their ancestors once fled from Mexico to Belize during the War of the Casts.

Although the national language is English, Spanish is widely spoken, as are English Creole and Spanish Creole. Mayan is spoken, too, especially by the older generation.

You will find lovely products like wooden carvings of sailboats, birds and bowls made of different woods that are typical of Belize. Beautiful black coral jewelry is made by gifted craftsmen. Even the cashew fruit is made into a very special cashew wine.

Interested? You can email me at helga.peham@chello.at.

I hope you will enjoy reading the life stories and experiences of the people who live here, many of whom have made their dreams come true.

PART I: BUSINESS

PART I, CHAPTER I:
AN AGING HIPPIE – PAMELA THE COOKIE LADY

"I was always somewhat a rebel throughout my life. In the '60s I was a hippie and I came to Belize looking for the aging hippies' home – and I found it."

STUDIES OF SOCIAL JUSTICE

Pamela the Cookie Lady was born in Portland, Oregon. She later moved to Seattle, where she studied at Seattle Community College. "I studied social justice and discovered there was none. I graduated with a Bachelor's degree in 1973 and moved to California."

In California she worked as a bank teller until, after a break-up with a boyfriend, she moved to Marin County, north of San Francisco in California, where she worked in an insurance company for a while. She spent the next 20 years in Marin County.

Pamela considers herself very fortunate in love, as she found two men who loved her.

"Larry was my first husband, then I was with George. Larry met George. He told George, 'I like to see Pamela happy.' He was great. Neither one of them ever came to Belize. They would have loved it." Both men died at the age of 51.

LEAVING THE U.S.

"I got completely disgusted with the political climate in the US in 1997. They were impeaching the president for having a girlfriend. So I decided, 'I'm leaving the States, this is not my home anymore.'

"I went to the library to check out guidebooks and discovered Belize. At this time I was working for a high tech company and I had stock options and money. I spent a three-week vacation in Belize in April 1997. When I returned, I sold everything, and moved to Belize six months later. On my holiday I had gone on a tour of Placencia, Corozal, Orange Walk, Caye Caulker and San Ignacio. So when I moved down here, I settled in San Ignacio."

SAN IGNACIO

"I moved to the Wild Wild West, to San Ignacio in Cayo District in the mountains. Unfortunately I experienced a lot of theft and corrupt police. They were stealing from me, and the cops were not doing anything about it. I lived in San Ignacio for three years and lost thousands of dollars. They broke into my house – a cement house with burglar bars and they could still get in! So I decided to move from San Ignacio to San Pedro."

ELECTIONS 1998

"Politics play an important role. Either you are red or blue. In 1997, the UDP (United Democratic Party) was still in power. 15% tax had to be paid, and the only thing they did was to pave the roads just before the election.

"Before the elections I was disgusted with the UDP, 15% tax and they were not doing anything! So I went to the PUP (People's United Party) and said, 'Put out a call for drivers to get people to the polls.' I offered my services. 'I have a car, and experience and my specialties are the aged, the infirm and the dead.' They said, 'You're hired.'

"On Election Day I was driving to the parade and a boy asked me, 'Can I ride with you?' He was the son of Dito Juan, the top politician and he had the largest PUP flag in the parade. So I had the biggest PUP flag on my car!"

"One of the biggest thieves in the UDP was Minister…"

COOKIES AND BANANA BREAD

Pamela came to San Pedro in 2000 just before Hurricane Keith, which took away everything she had.

"I was very poor at this point. I didn't know what to do. My Belizean friends advised me to 'go out and sell.' 'I don't know what I can sell,' I said. My mother always made good cookies, so I thought, 'I suppose I can make some good cookies.' That is when I reinvented myself as the Cookie Lady. I've been selling cookies ever since.

"I would like to say that Belizean people – although this is a poor country and there is petty theft all the time – have been the most welcoming people to me and have made me feel at home. 'Come on, go out and sell.' They gave me good advice on how to live in this country. I have many Belizean friends. They are the ones I count dearest.

"I sell the American classics: oatmeal, raisins, a lot of chocolate chip – and banana bread. I checked out a book from the library and then tinkered with them until they came out the way I like them, namely how my mother made them. I sold them in the street – I am a street vendor. I also mostly go down to the beach. It took me a long time to get Belizeans to try my cookies and bread. They don't know what cookies are. I bake at noon, when it is already hot. I come out around 4 pm and sell until all the cookies and the banana bread are gone. September and October are real slow months. I am learning to tuck some money away. Most of my customers are tourists, although I also sell to the local expats.

"I sell to Captain Morgan's store. Captain Morgan has a little convenience store. They send a boat every now and then, and I live close to the chicken store, so they pick up my cookies and banana bread when they come to pick up the meat and eggs from the chicken store.

"I sometimes have tourists say to me, 'Do you remember me?', 'You look kind of familiar to me,' or 'I bought cookies from you three years ago.'

"I'm only 58. I may get social security much later. I am happy in Belize, I support myself, I am a little micro enterprise and it supports me. I am a resident."

PAMELA'S ADVICE

"If you move down to Belize, remember: Belize is a developing country with a small population. Personally I would like it to stay that way.

"But if you are a person who is retiring, this would be a wonderful place to spend your retirement. On the other hand, if you have a business idea that is not damaging, you will be welcome in Belize.

"This is a country of many peoples, even though it is a small country – not as big as a small town in the States. There are many Mestizos, Creoles, Garifuna, three different kinds of Mayas (the Yucateco, the Quechi, and the Mopan), and the German and Russian Mennonites and many expatriates: Canadians, Americans and some Brits. There are also the immigrants escaping civil wars in Central America: immigrants from El Salvador, Honduras and Guatemala, but people get along. It is a rainbow country.

"To paraphrase Kennedy: If you come to Belize, don't ask what Belize can do for you, ask what you can do for Belize. And for my part I am selling baked goods!"

PART I, CHAPTER 2:
THE BIRD AND HIS WINGS

John Greif III is the owner and president of Tropic Air, one of two airlines operating within Belize. His father is American and his mother is Belizean.

JOHN'S CHILDHOOD AND YOUTH

John was born in Kentucky to a middle-class family. His father was a flight instructor and former World War II pilot who met John's mother in Belize City. His parents moved to Belize when John was five years old, and John's paternal grandmother raised him in the States. "I spent my life in and out of Belize. During school breaks I often went to Belize. When I was 19, I moved to Belize full time."

John graduated from the US public high school system and enrolled in a program in aerospace engineering in Texas. After two years John decided that engineering was not for him, and he came to Belize without completing his college degree.

BELIZE IN HIS YOUTH

The Belize of John's childhood was a rustic country. When his father came in 1962, there were no hotels, no electricity and no internal combustion engines. The flight instructor was living in South Florida when a former student, Kurt Bender, owner of the airline "British Honduras" and a former Luftwaffe glider pilot, contacted him about working in Belize. They'd flown on opposite sides during World War II but had nonetheless become good friends.

John's parents started Holiday Hotel, and his mother took over ownership of this first hotel on the island while his father founded a charter flight business. (John's father later wrote down, in a book that was distributed only within the family, a number of stories about Belize.)

Send me an email at helga.peham@chello.at if you want to contact me.

"My memories of Belize City of that time are dim," John says. "San Pedro was like a big playground, very socialized. Everyone was your parent. I played with everyone. You ate everywhere you were. It was like a utopia in the late '60s. We went to school in Kentucky and passed our holidays here."

MOVING TO SAN PEDRO AND THE EARLY YEARS IN THE CHARTER BUSINESS

John decided to move to San Pedro in 1978, when his parents were living on the island of Ambergris Caye. John was a young man, fresh from school and looking for a career. He worked in construction for a while, then spent a year on the water, contemplating life as a sailboat charter captain. Eventually it was the skies that called to him, and he decided to enter the aviation business like his father. "My father being a flight instructor, I was pushed into aviation all my life – he was pushing and I was pushing back."

John realized there might be opportunity in Belize City, where there was only one airline – Maya Airways – and only one charter operator. John invested in a three-passenger airplane, a Cessna 172, in 1979. The Tropic Air flight service they offered was almost an instant success. The next year, the family hired another pilot, and three years later they bought a second plane and applied for a scheduling license. They also bought a brand new Cessna C207, adding a C207 every year for the next four to five years, with John doing most of the flying.

Tropic Air was the first airline in Belize to sell individual seats. John is convinced that Tropic Air's reliability is the key to its success. He believes that if you seek only money in life, you don't get it. "In life, the thing you seek the most you never attain, and business is part of life. Right from the beginning, we were driven to do a good job. That is the reason why we were and are profitable."

When John first started flying, only a few international

tourists came to San Pedro, then just a small town with an airstrip. The only flight to the United States was one by Taca Airways that left Belize City in the morning and returned in the afternoon. In the morning, John shuttled passengers from San Pedro to the airport in Belize City, where he would wait all day for the Taca plane to land. He was not allowed access to the arrival areas in Belize City, so he depended on porters to bring the passengers to him.

For most tourists it is their first ride on a small airplane. "Once I had a very nervous couple," John recollects. "The plane had just a few seats, I was young, and that day the weather was bad. 'Ma'am,' I assured the lady, 'I have done this for a long time, and I haven't ever left one up there.' She smiled and boarded the airplane.

"Another day a porter had brought an older American lady to the airplane. We were waiting together by the airplane while the porter went back to find the rest of my passengers. I began to talk to the lady. She said, 'Young man, are you waiting for the pilot too?' 'Ma'am,' I said, 'I have some very bad news for you. I *am* the pilot.'"

PART I, CHAPTER 3:
YOUNG GRINGO LOST AND FOUND HER HEART IN BELIZE

Dara Jones owns a shop in San Pedro selling old and new clothes from the US, mainly to locals. The author (helga.peham@chello.at) talked to her on the terrace of Sam's hotel right on the beach on a lovely evening full of moonlight. She has a heart-warming story to tell about how she met her Belizean husband and settled down.

A CHANGE IN LIFE

"I decided I had to get my life together. I worked at Continental Airlines in International reservations in Denver, Colorado from 1994 to 1998. I was searching for a purpose and a meaning to my life. I was 21 years old. Life was really going nowhere.

"I moved with my family from Kansas to Colorado when I was thirteen and got into the wrong crowd of friends. I was introduced to the teenage life overnight, from being a country girl to a city girl. I had boyfriends for 1 to 1½ years, but I never had a good boyfriend, or a good experience with men.

"By the age of 19, I was living with someone who was 13 years older than me. I was doing everything a wife would do, cooking, cleaning, but we were not married. I found out he was an alcoholic and he didn't treat me well. So I started to think, why am I here? I was searching for spirituality and the purpose and meaning of my life.

"I went to college at night part time for a little under two years and worked full time. Good marks were the result. I got a 4.0. When I went to college, I knew there was more to life than what I was doing. I moved back to my parents' house. That was like closing one chapter of my life and opening another one hoping for something better.

"Not believing in God I nevertheless was trying to understand who God was and what Christianity was about. I just started praying on my own, testing of Him if He was real. And that started when planning a trip to Belize to relax and get my thoughts together, wondering what I would do next."

THE TRIP TO BELIZE THAT CHANGED HER

Dara had been planning to go to Belize with some girlfriends from work, but a couple of days before the trip, all of her girlfriends cancelled.

"I booked my trip to Belize and began to pray. 'What will I do,' I asked God, 'in Belize, without a companion?' When I was checking the flight manifest on the computer of the airline I was working with, God impressed a name of a passenger, Christy, who was a flight attendant from Newark, New Jersey. Reading her name on the computer screen, I felt like God was telling me that when my flight got to Houston he would show me who this lady was and we would share the flight, the hotel and the expenses.

"Keep in mind, I was not a Christian, I was testing, I wanted to see if He was real. I didn't have any faith in my prayers and I didn't know that He would do anything for me but I knew I didn't have anything to lose by trying.

"So when I arrived in Houston the Lord impressed me again. A young lady stood in the passenger hall wearing a tie-dye T-shirt and cut-off Levi's jeans shorts. There were at least 200 people, maybe more. There was no way I could have found this woman on my own. It was in terminal C in IAH [George Bush International Airport], Houston. I laughed to myself when God showed me the young lady, because anyone who has worked for the airlines knows that you must have professional casual dress when flying, even in your spare time.

"I prayed to God to show me someone else. 'This is not the person, she could not wear these clothes, I am not so foolish as to believe that.' He didn't show me anyone else. So I said, 'What shall I do now? Let me go ask her.' I said, 'Excuse me, is your name Christy X?' She said, 'How did you know my name?' I said, "Don't worry about that. You are traveling alone. I'd like to know if you'd like to share a hotel room with me.' She was also looking for someone to share a room with. We ended up paying $20 each to fly first class, and getting to know each other over our filet mignon dinner."

They arrived in Belize and flew to San Pedro on Tropic Air. When they landed they walked around on the sandy streets, pulling their suitcases, looking for an inexpensive hotel. They ended up at Lilly's at the beach, where they rented a room upstairs.

"I don't remember much about my trip except meeting the

man who one day became my husband, but that came later.

"I remember snorkeling, seeing the nurse sharks with my underwater camera and making a getaway to the boat, because I was frightened. I remember lying on the beach. I remember waking early and taking our version of Sports Illustrated swimsuit pictures on the beach. We climbed on someone's yacht on the ladder, and took pictures like it was our boat. We went up to the cabanas at Ramon's Village, onto the steps where someone was sleeping behind the doors. We took shots as if they were our own cabanas. We climbed palm trees, lay down, fooled around, played in the sun and had fun. I had met a new friend."

FOUR KIND POLICEMEN AND A LOST KEY

"One night we went out to dinner. When we were leaving the restaurant we met three policemen who sat at the bar talking with the bartender. We started to share our driver's license photos and passport pictures, and talk about the places they were from – just friendly, getting to know each other's cultures and backgrounds. We told them we were going to Big Daddy's Disco. The young policemen asked if Christy and I would like an escort because they were going there as well. We agreed and we all headed out to Big Daddy's for a fun evening.

"Towards the end of the night another policeman walked in to get the key to his hotel room. He was tall, dark and extremely handsome. He was Creole, a mixture of black, Spanish and a bit of Indian. I was immediately attracted to this man and I wanted to know who he was, though coming from the life I had been living I knew I was beginning a new chapter, and I wanted to be careful to think about the things I was doing. I didn't want to start a relationship based on physical attraction.

"All he wanted to do was to get his key and leave. I invited him to stay for a drink and he didn't want me to buy him a drink. I tried to talk to him. I tried to buy him a drink, I asked him to dance and none of my efforts were fruitful. He didn't want anything to do with our party, our fiesta.

"A little later it was my turn to buy drinks. I brought him a drink and he refused it. I was quite offended and told him he could either have the drink or be extremely rude and leave it to waste. 'You can drink it or really upset me and be rude.' After a long while he started to sip on his drink. Then we started to talk a little, not much. Again I asked John to dance. And he finally said, 'Ok, ok I do one dance.' So he started to dance. And we did not leave the dance floor until Big Daddy's closed. We had a lot of fun, he and I conversing and dancing together."

After Big Daddy's closed, the group went walking along the beach. The sea was inviting, and they decided to take a late night swim.

"As you can imagine John didn't want to have anything to do with this adventure either, so he continued walking on the beach while Christy and I and the three men jumped in. John was walking on the beach, looking to walk away. I immediately became concerned about losing my new friend. So I began asking the other men how to say something in Creole to John to try and sweeten him up, make him hang around a little longer. They told me to say, 'Con ya John.' I began to holler it several times and he looked back to see who was speaking to him. He was surprised and tickled to hear a white girl beckoning him to stay."

Later, they realized that someone had stolen one rather chubby policeman's pants as a joke. He was left to stroll on the beach wearing only his boxer shorts. Unfortunately, the missing pants contained a key to one of the two policemen's hotel rooms.

It was about 2 or 3 in the morning. The small party walked around the island but could not find the owner of the hotel. The four policemen from Belize City had one hotel room with two single beds. Two policemen went to their hotel while John, Robert, Christy and Dara went back to Lilly's to sit on the veranda and chat.

"I was a thrifty traveler and often traveled with barely enough money for the trip. This was common among airline employees with $20 tickets.

"John and I sat talking on the veranda and I pulled out my Vienna sausages, my Twizzlers (red licorice), and my crackers, and we began sharing my junk food. I always carried these staples in my bag in case I didn't have enough money to buy food or did not like the food in whatever country I visited. We began to share about ourselves and our families, and we started getting to know each other. Many years later John said that it was at this point that he started to fall in love with me.

"The time passed and Christy and I spoke and agreed that we would invite them to stay the night with us so they could sleep in a bed. The men were part of the Dragon Unit of the police force and they were leaving early in the morning by boat to return to Belize City.

"They accepted our invitation. Robert was delighted to be sharing a bed with Christy while John seemed a bit reluctant about the whole idea. I told John to make himself comfortable, feel at home. I wore my night gown, laid down, covered with a sheet and John deposited himself on top of the sheet fully clothed with shoes and all. I asked him, 'What are you doing? Aren't you going to get ready for bed?' He replied, 'I am ready for bed.' I said, 'How are you getting any rest sleeping in your clothes and shoes?' I told him he could at least take off his shoes and jeans, and try to get comfortable enough to sleep for the night. He reluctantly removed his shoes and jeans and lay on top of my sheet with his socks, boxers and T-shirt.

"All the while Robert was in the bed next to us touching Christy every opportunity he got. Christy was continuously telling him to keep his hands off her and to leave her alone. John told Robert if he didn't respect the lady he would kick him out of the room to sleep on the beach. Robert finally gave up and realized that Christy was not interested in him. The next morning she confided to me that she was also attracted to John.

"The following day I awoke to find myself under the sheet, just like it was when I went to sleep. John was sitting on the edge of the bed fully dressed, patiently waiting for me to awake. I asked what he was doing and he told me he just wanted to say 'goodbye' and thank me before he left. We all went out on the veranda and exchanged addresses and phone numbers to be pen pals. We gave our hugs, said goodbye and the two men left. I told John that I would be praying (to this God I did not yet know) that they had a safe trip back to Belize City.

"I told Christy that there was something about John. I wasn't sure why, I couldn't put my finger on it, but I really didn't want him to leave. So I asked her to go with me down the beach to the police station so that we could see John once again before he left. Christy agreed and we ran along the beach to the police station. Many of the police officers were already in the boat."

These men were part of the police force's Dragon Unit for serious crime. John should have been on a week-long jungle patrol but was called off the patrol the last minute to come to San Pedro because a gathering of ambassadors on the island was short of security. About 20 officers came on a small boat from Belize City. Their boat had broken down several times on the way out to the island and none of them had lifejackets. Unfortunately the Belizean police force is often short of funds to acquire the necessary equipment for their work.

"All of a sudden John walked out of the police station in his camouflage police uniform with an M16 strapped over his shoulder. I was shocked. I did not expect to see him with this sort of weapon. I quickly realized the seriousness of the line of work he was in. He was surprised to see us. I told him I enjoyed meeting him and I just wanted to let him know that I really wanted to see him again. He said he would appreciate that and that we would keep in touch. I leaned over the bench and planted a kiss on the side of his cheek. All of his fellow officers immediately began to tease and harass him."

Dara phoned him later in the day.

"When he answered the phone he did not even know who I was. It took a couple of minutes for him to wake up and realize who I was. He was tired from the trip and had been taking a nap. He told me that they had got back to Belize safely without any breakdowns or problems. This was one more answer to a prayer. We chatted for a bit and agreed to talk the next day."

A CITY TOUR AND A SURPRISE TRIP

"The next day when I called he invited Christy and me to come the following day to Belize City to spend some time with him and his cousin before our flight home. The two men wanted to take us around the city and we agreed.

"At the last minute Christy cancelled, so it was only me joining John and his cousin in Belize City. They had rented a Suzuki Samurai to take me around the city. They showed me a few sights before John explained that they could not travel around as freely as he wanted because it could be dangerous if someone recognized him. John decided to take me to a surprise place. Twenty minutes later, we arrived at the Belize Zoo. I got to see the wonderful animals in their natural environment. It was a wonderful first date.

"Then it was time for me to head back to the airport. In the departure hall, John told me that his mother had invited me to come to Cayo to see more of the sights his country had to offer. I told him that sounded great and that I would be back next month. He said, 'Next month?' I said, 'Why, is that too soon?' He said, 'No, I just did not expect you to come back so soon.' I explained to him that I could visit for the price of a phone call. We hugged goodbye. I planted a quick kiss on his lips. Again he was shocked and I thought, 'What's new?'"

They began to talk via phone over the next two weeks and got to know each other better every day.

DEEP FAITH

"The following month, in November, I accepted Christ as my savior in Colorado at my mother's friend's house. I called John to explain that my life had changed that day. He told me that he was also a Christian and was excited for me. I thought when I called John to share the news, it could be the end of our friendship. Instead, it was the beginning."

A RELATIONSHIP GETS SERIOUS

Dara visited Belize several times in the next few months.

"One day when we were traveling to the airport in the back of a pickup truck John said to me, 'You know this isn't a game.' I said, 'Yes I know, but what do we do about it?' He said, 'I don't want either of us to get hurt.' It was at this point that we both realized how serious our relationship was and how strong our feelings for each other were becoming.

"In 1995, John came to Colorado and worked as a driver for UPS. He lived in my parents' spare bedroom until October 1995 when we married, one year after we had met."

The couple has two children. Janae was born in Houston

and Jasmine was born in Colorado. From 1995 until 2004 they traveled back and forth between the US and Belize frequently. For many years they had US $20 flights, and later, John learned to travel the roads to Belize through Mexico.

MOVING TO BELIZE

"It took me ten years to let go of the United States and come to Belize, although my heart's desire was always to move here. We bought a 24-foot moving truck on Ebay for US $3,000 and we packed everything we owned into it and hit the road.

"We opened a small clothing shop in San Pedro. Our shop is on Middle Street in town. At night we would move the clothing racks forward, blow up the air mattresses on the floor. We have a shower in the shop but no kitchen facilities, so we ate out every meal.

"One year later we decided that we were fed up with paying rent and we wanted a piece of land. We had not a penny in the bank to do this, so we had one of the island's largest yard sales. We sold everything we owned to start saving money for our land. In October 2005 this dream became reality and we bought a piece of land on the north side of the island.

"In June of 2005, John took a job in the oil industry in New Mexico, where he works two weeks and has the next two weeks off. Now we all stay together in a hotel in New Mexico and on our two weeks off we all drive to Belize with some inventory for the store. We spend four days driving down, a week in Belize and three days driving back up. It's a crazy life right now, but we know it will be worth the sacrifice. We are hoping to do this for only two years. I home school my elder daughter so that we are able to travel as needed. And oddly enough though the lifestyle is crazy to many it has blessed us beyond our wildest dreams. We have a wonderful time doing it.

"While we are away, Michael, our shopkeeper, lives at the shop and takes care of the daily business.

"The slower pace here helps me to relax and to realize what is important in life. It helps me to keep my focus on our family. People here are wonderful. We fit in with everyone. I feel like I am at home, like I am more comfortable here than in the States."

PART I, CHAPTER 4:
FROM NEWS TO FROZEN CUSTARD

Dan and Eileen Jamison published the San Pedro Sun from 1997 to 2004, after moving to the island in 1996. They now operate a food business. Eileen and Dan both grew up in Meadville, Pennsylvania. They like the fact that Belize is an English-speaking country, and also that you can bring a dog without quarantine.

MARRIAGE IN MOOREA

Eileen stands behind the counter in their shop and vividly tells me her story, assisted by her husband Dan.

"I was a country girl. I grew up playing in the woods, riding horses and that kind of thing. Dan was a city boy in the suburbs.

"I was a receptionist in my first job. I did a lot of different jobs. I was an office clerk, an automotive re-conditioner. I worked for a car dealership. I would recondition used cars, and they would sell them. I was the first woman to do that. Women are better at details. I worked in a deli bakery, I made doughnuts, I did flower delivery and I was a bartender and a bar manager.

"The first time I met my husband was when I backed into his car, which was a brand new car. I broke his grill. So I left a note on his windshield. I thought he was cute. I did not meet him again for ten years. Dan was managing his parents' beer business. I was managing a bar. We first met around 1976 and again in July 1986. One day he walked into my bar and I recognized him. We ended up going out as friends at first and then we traveled to Europe together in March 1987. By mid 87 we moved in together.

"After I met Dan I worked in an outpatient intravenous pharmacy. We did intravenous feedings in people's homes, pain control for patients who had cancer. I was a pharmacy technician, so I compounded the drugs, did all the purchasing and the inventory. I did that for eight years. It was a great job. People appreciated my work.

"We had been traveling for eight years in the Caribbean. Every year we went to the Caribbean and the South Pacific, looking for a nice tropical island to live on."

In 1994 they spent a month in Australia. Eileen and Dan were on their veranda at a "Bed and Breakfast" in Melbourne when they learned they'd received approval to marry in Moorea, in French Polynesia.

"We had a traditional Polynesian wedding; we had a legal ceremony in the Mayor's office before. Dan was tattooed from 'forehead to foreskin.' It was gorgeous. They took him to a separate island and tattooed him with something like a permanent marker, with Polynesian symbols.

"They stripped me, massaged me with oils, and redressed me in a traditional dress. I had four bridesmaids and a maid of honor that I didn't know. Women brought me to the beach to meet my husband-to-be and men brought him by canoe. They announced his arrival by blowing on a conch shell. It was really a magical moment. A big-bellied Polynesian priest presided over the ceremony. A man on horseback translated the ceremony into English.

"We had a great feast after the ceremony. They picked us up in one of the palm thrones. They carried us over their shoulders through the village. They had a reception for us. Then we had our feast and the whole village participated. It was a joyful event. We drank champagne from coconut halves. Then they took us on a sunset boat cruise in a canoe. It was great fun."

MOVING TO BELIZE

One and a half years later they traveled again to the Caribbean.

"We learned about Belize and cancelled a trip to Greece and Egypt to come down here instead, in December 1995. We did a whole country tour from Corozal to Punta Gorda, also to San Ignacio. We went to Caye Caulker. San Pedro was our first stop, Victoria House the best stay, with the nicest beach. We were two weeks in Belize, went to Caye Caulker, then returned to San Pedro and stayed an extra week here.

"We moved to San Pedro in December 1996. We had sold everything, except our house. We sold it only a year later."

THE SAN PEDRO SUN

They soon found their business opportunity on the island.

"In February 1997 we were offered ownership of the newspaper San Pedro Sun. This newspaper was a great learning experience and an adventure. The reason we wanted it was that we wanted to give a voice to the people of this community. At the time a lot of people were subservient to the government. There were not many people who spoke up. People were treated unfairly. Some of the utilities were unreliable. It is not easy living here all the time."

They learned how to use a computer only three months before they moved to San Pedro. After a one-month training period from the former owners, Dan and Eileen were on their own with the San Pedro Sun newspaper.

During that time Eileen helped found the Saga Society, in March 1998. Its goal is to promote kindness and to maintain control of the dog and cat population on the island. They used the newspaper as a vehicle to promote the society.

Their one employee, Felix, was the associate editor. The three managed to publish a newspaper every Thursday. When Hurricane Keith struck in 2000, the paper was out ten days later. The San Pedro Sun had to move their computers downtown to Cannibal's Restaurant because there was no electricity in their area.

HURRICANE

"Contrary to rumor, our death was greatly exaggerated. CNN said that the whole island was wiped out. This was an irresponsible news report, very upsetting, greatly exaggerated. We used the first couple of issues after the hurricane to prove that the devastation was not as widespread as media in the US made believe. Tourism is very important to the island. The islanders rebuilt and tourism was generated again. Official agencies denied that there were deaths. There were dead, but not confirmed. They would not let us have access to confirm or deny it. At the time we published the only 'gringo-owned' newspaper. It was of a different standard and the only independent newspaper. Most other newspapers were politically owned or influenced at that time."

NEWSPAPER REPORTING

Dan and Eileen got their stories by attending all the community activities and town meetings, and hitting the streets.

"We quickly gained the trust of the locals. They would come and offer us confidential information to help our investigative reporting. Our delivery day was Thursday, even if there was a holiday on Monday or Tuesday. Thursday was Thursday and not a day later. It is a responsible job and you take that responsibility. People sometimes don't come to work when it is raining, here on this island."

The newspaper was an important learning experience. "We now know and love our community better, despite the dirty little secrets."

They introduced people of the community to the world in the 'Isla Bonita Our Community' column.

"We had a lot of positive feedback from the column. We had the largest group of foreign subscribers of any newspaper in the country. Our newspaper went to six different countries at one time. Our website had four times as many hits as any other Belizean news site."

They had a five-year plan for the newspaper and they sold it after seven years. Once they decided to sell the newspaper, they advertised it on their website. They turned down a few people who might not have had the community's interests at heart. "We turned three people down who had the money on hand. We sold the newspaper business in September 2004."

FROZEN CUSTARD ON MIDDLE STREET

After selling the newspaper, the couple took a year off to rest and rejuvenate. They then made a big move, entering a completely different line of business, one that ensured them constant smiles. Dan and Eileen opened a frozen custard (ice cream) store on Middle Street in San Pedro.

"I love my job, this is my dream job," Eileen explains. "People are always happy. If they're not happy when they get here, they're happy when they leave. We grew up with frozen custard in Pennsylvania. We never really had sorbet, a frozen fruit dessert, until we made it here. It is used to cleanse the palate between dinner courses and wine tastings and cheese tastings. We use it as a frozen dessert. It has no dairy in it. It's a big chunk of our business."

Dan remembers how everything started. "The first step was that I wanted a really good chocolate milkshake. And I remembered the ones I used to get in the States, and Hank's back home used to make frozen custards. The only way to get it in San Pedro was to open our own frozen custard shop. So after investigating this idea I went to the US to learn how to do it and to find the equipment. I went there for training and buying. And, because we were off for a year after we sold the paper, I had the time. And that was basically it. We saw the need on the island. Nobody in the country made frozen custard, so we thought this was a little niche, something unique. We weren't competing with anybody. This was a completely different product; it is gourmet ice cream."

They opened up officially in September 2005. Before opening the shop they gave out their frozen custard for free to test if people liked it. It has been a hit on the island.

"Most tourists come in every evening," Eileen explains. "The locals love it. It is nice to have a business where everybody comes in happy and leaves happy."

They keep their shop open from 2 pm until 10 pm every day except Tuesday. During high season they open half a day on Tuesday, if the mood strikes them. Tuesday is their day off. "We are not greedy, we never were greedy. We want to put out a quality product at a decent price, the same as we did with the paper."

They'd like to keep the business small and casual. "We want to be able to take time to enjoy where we live."

GOOD ADVICE

"I would say to live here, your needs must be simple. If you don't expect much you will be pleasantly surprised; if you expect a lot you will be sorely disappointed. Keep your needs simple and don't be greedy. Life will be beautiful here and you will like it. And you will be accepted that way too. It is a wonderful place to live, if you just let it flow along like the tide. It all balances out in the end."

For success in business, they suggest: "Please be inventive, don't do what everybody else has done, is doing. And definitely research, do your homework, before you jump into a business. And when you have a bad day, custard makes everything better. Having this shop is like a bar, but you don't have all the drunks; you have all the nice people."

The couple has made many friends on the island. "Lions is a good organization and in San Pedro, very active. It is just one of the best organizations the island has ever had. They provide for almost any type of need: food, clothing, medical, financial, building, etc., and do a really good job at it. There are hundreds of members of the Lions Club, Belizeans and expatriates. There are five different Lions Clubs, the most recent one in Caye Caulker.

"We like it here. It is not for everybody. That is why there are so many places in the world to live. But if it is for you, San Pedro is a good place to live."

PART I, CHAPTER 5: ARCHAEOLOGY AND GOURMET FOOD

William Chappell Ross was born in Mercedes, Texas, where he grew up on a farm. He has worked in restaurants throughout his life and is now the manager of the restaurant Stained Glass Pub in San Pedro on Front Street opposite the Belize Bank.

DIRECTOR OF ARCHAEOLOGICAL CAMP IN BELIZE

In his thirties, William resumed his university studies and obtained a degree in art history and anthropology. During that time he came to visit Belize on an archaeological expedition. William was engaged in the Belize Archaeological Project and Program and worked for this program as a student for two years. From 1996 to 1998 he worked as a camp director for the same project while still doing archaeological surveys.

CHAN CHICH LODGE

William moved to Belize on a full-time basis in 1998. He became a chef at Chan Chich Lodge. With an archeological site nearby he was also able to keep digging.

"Working in Chan Chich was wonderful. I love the jungle. I got to meet amazing people. Laura Bush was here with her high school girlfriend, just before George W. Bush announced that he was running for president. It was really interesting to meet Laura Bush and her mother. Her mother was the stereotype of a Southern woman from the United States. Very proper, incredibly sweet and friendly – she was an amazing woman to me.

"It was pretty much day-to-day life. I got up at 5:30 am, made breakfast for 2 to 4 hours, then prepared dinner for the visitors. Interacting with the guests was always fun – that was the most interesting part. Everybody from fashion designers to the first lady of the US was there." Numerous biologists and archaeologists also stay at the Lodge.

"I was blessed by being able to see four different jaguars in the wild. Once I walked on trails and saw two jaguars. They were the spotted ones. Once driving to the zoo, I saw a black one and another spotted jaguar. My sister and I were driving back from Lamanai and coming through the cleared portion of Gallon Jug when we got to watch a jaguar stalking a deer. That was amazing."

While working in Chan Chich William got to know the kindness of people from San Pedro.

"We were stranded there during Hurricane Mitch, guests couldn't get out, staff were at home with their families. I stayed there. San Pedro was evacuated. We had people from the island of San Pedro washing dishes, as waitresses, actually a whole lot of work and a whole lot of fun. Dian Campbell waited tables and a local realtor, Jackie De Vine, helped me cook. The owner of the only gay Bed and Breakfast on the island washed dishes for me. She was the most loved expatriate on the island. Later I moved to San Pedro because of these people. We had a management change, so I changed. When the managers left Chan Chich, I left. I felt I needed to move on at that point."

William moved to San Pedro on the island of Ambergris Caye in 2001.

SAN PEDRO CALLED THE CHEF

Before joining the Stained Glass Pub, William worked at the Hideaway Hotel, then the Blue Water Grill, where he learned to cook Pan-Asian style cuisine, and then Banyan Bay.

"At that point I was offered the position of manager at the Caribe Island Resort. I ran that restaurant until the condo owners got into a major fight and closed the restaurant. Then I was at Sundivers and El Pascador. Finally, I was offered this job, this opportunity at the Stained Glass Pub."

William also helped the owner of the first coffee bar in San Pedro to establish her bar with delicious espresso, coffee and other cocktails, and has acted as chef at the Tackle Box, the well-known disco and dining establishment. William loves new challenges.

As for his personal life he says, "I am single, gay and happy. I love snorkeling, hiking, bird-watching."

PREPARATION IS NEEDED

"I would recommend that every expatriate does long and deep research. Don't jump into anything, but do research. On the other hand, Belize is my home and I'm not leaving."

PART I, CHAPTER 6:
HOTEL MANAGEMENT AND BUSINESS RECOMMENDATIONS

Sheila Nale is the Manager of the Mayan Princess Hotel. Her husband, Rusty, is responsible for the maintenance of the building and the condos there. They moved from the USA to Guatemala and Central America in 1989.

Shortly before that, Rusty's father died. "That was what started us thinking," Sheila recollects. Rusty had plans for what he wanted to do when he retired, but he decided not to wait until he was an old man to realize his dreams, because then it might not ever happen. Rusty was sure they needed to act on those dreams while they were still young.

"We heard that Belize was an English-speaking country, up and coming religion wise." Sheila and Rusty are Jehovah Witnesses. "We wanted to volunteer, live there, be of help. We thought we had to check it out."

A YEAR IN GUATEMALA

They planned to check out Belize from Guatemala, which is where Sheila's parents lived, but they stayed in Guatemala for one year instead, coming to Belize in 1991.

Rusty had previously been working in building and maintenance services, so he was well-suited for hotel maintenance work. Sheila, however, had been a hairdresser for more than 10 years in the US.

"How I became a hotel manager from a hairdresser, I don't know. You stay in Belize, you meet various types of people and find that you have opportunities to do what you have not thought yourself ever capable of."

BELIZE CITY

Sheila lived in Belize City, in Ladyville, for three or four years. She first worked in Belize City as a hairdresser while Rusty did many different types of work. For a while he worked in construction. Then he and an associate brought in food from Guatemala, with Rusty overseeing the operation.

"When you live in a Latin American country you find out that Belizeans are very fast and efficient by comparison," says Sheila. "I remember the first time we came to Belize by road. At the border, the street went from gravel road to paved road. We got on the bus. Everyone smells good, smells are different. And we went from that rough rocky road onto a paved road. A move up, we thought, faster. We felt like everything was very efficient. Everyone should go to Guatemala first before moving to Belize."

MANAGER OF A JUNGLE LODGE

They next lived outside of Orange Walk for a year, in a jungle lodge.

"We managed a lodge in Orange Walk with 18 cabanas and various support buildings, on the lagoon, a real jungle lodge, for people who like bird-watching, jungle hikes and the Mayan ruins. Rusty and I learned how to run a jungle lodge.

"We were assistant managers and it was a very good and quick introduction to the hotel business. You had to do everything: water, electricity, airstrip, meals, manage a big staff."

MOVING TO SAN PEDRO

After that they came to San Pedro on Ambergris Caye. When they first arrived in San Pedro, they worked for a company that deals in rental properties, Key Management, and Sheila also did some hairdressing. Within six months, the Mayan Princess became available. The former managers were business acquaintances of the lodge.

MAYAN PRINCESS

"The jungle lodge was fun, but very consuming. You need to have a life as well. We had to decide to take a little of our life back. We had skills. This worked well for us. We worked full hours and managed the hotel."

MAYANS AND GARIFUNAS

"Mayans and Garifunas have their old beliefs mixed in with Christianity. They have retained much of their old thoughts and fused them with some of Christianity. You find Mayans who attend church, but when time comes to plant the cornfield they say prayers to the four corners, they say prayers to the various gods when they chop the field before they plant. When they cut first fruits they usually offer some to the corn god.

"The same applies to Garifunas. Most are Christians. Still, they take their traditions seriously, such as doing the dugu, also called Feasting the Dead. They still believe strongly in this. If someone dies, this is a seven-day thing, all tied up with the spirits. There are tons of superstitions in Belize. Many people live by signs; they take everything to be a sign. Most babies have red bracelets to protect them against evil."

GOOD BUSINESS POSSIBILITIES AND PRINCIPLES

If you want a change, Sheila recommends, "Do it while you are young enough to enjoy it. Don't put everything off.

"Make sure you check carefully before you buy property. Never buy anything in Belize that you haven't seen in daylight and night, sunshine and rain, and be sure you know the neighbors. Never invest your money unless you are completely well informed about everything. Some people don't do that. They come on a wonderful vacation, fall in love with the place, a condo, a house, see a seagull, pull out their checkbook and buy, an impulse purchase. They should also come when the tide is high. Electricity is sometimes out. Stay at least three months. Just come out here, rent an apartment, and only then look for property.

"There are many business possibilities. People need to do their research before they move. You will have more success, an easier time, if you start a business where you are going to hire Belizeans. Look and see what is lacking. Remember that certain kinds of business are protected, like tourism. You cannot get a tour guide license if you are not Belizean. Before you get financially and emotionally engaged, talk to the Belize Tourist Board. They can advise. They want investment in Belize. But they do not want foreigners to take jobs from Belizeans, but to provide jobs for Belizeans.

"To open a business is not so difficult. The Chamber of Commerce or Tourist Board can give you advice on what would be viable.

"In San Pedro there are quite a few businesses. Even a bookstore does OK. Local people like that bookstore. You need to come, be part of the community, and find out what you are missing, what you would like to have. Ask the community; ask the people, what do you have that they would buy? There is the Sausage Factory in San Pablo, a brilliant idea. It was recently sold. Then there is Caye Coffee. They sell Guatemalan coffee, they roast it. Odd little things that you think of are business opportunities. These were necessary businesses. It took someone from outside to fill that need, like the sausage factory. The owner of the sausage factory was an American. He built the business and sold it. It is clean, produces and sells excellent products. Also restaurants, hotels, quality goods that they don't have to import are an opportunity. Study the community. What can I do to fit in here? You can establish a business and then sell it. If you have a business to sell just list it with realtors, or sell it directly. Put out the word. Michael Fox, CPA US, has a business exchange. If there is an area you are good at just buy a business. Walk in and ask if you can buy it."

LAND DEVELOPMENT

There is a lot of land purchasing still going on. Sheila doesn't suggest that you buy a condo as an investment, but rather as a vacation home.

"A condo is perfect as a vacation home. If you want to invest, buy and hold on to a property, then resell. Raw land is probably best. Here land seems quite cheap. Many people are investing. Belize features international home hunters. It seems like land prices are going up astronomically here.

"The Northern Island of Ambergris Caye is developing fast. There are huge developments. The bridge is open to the Northern Island. If they build roads, things will get big. Sueño del Mar is the first of its kind – fractional ownership. This is new in Belize, very luxurious. You buy two-month sections. Each unit might have six owners.

"If people come here to the island and the weather is bad, people buy. They go to real estate companies and look at property. The tourists cannot go out diving, so they say, 'Let's go out and look at properties.'"

THE ISLAND'S FUTURE

"If managed right, you get a boost in tourism. It could grow wildly. It is necessary to work on the infrastructure, the roads, and it is necessary to control the growth. This is difficult because of politics. Political feelings are so strong, often different to local governments. It is bad not to have a non-political forum to put forward good goals without political parties and government. Infrastructure is the biggest thing — roads and all the support services like electricity, water, everything, even the controls, licensing guides, captains of boats. Make sure tourists are safe and be able to provide good product. Tourists have become more demanding. They want a high level of service and they want the area to be pristine."

BEAUTIFUL HARD LIFE

"There are a few things people should know when they come here: You cannot feel the need to control everything. Here you find how little you can control. Americans who have that need to control everything in life find it very stressful here. If you have a vehicle, a golf cart, you are not protected from the elements. You need to be driving in your raincoat.

"Things have greatly improved, but still, say, you have a craving for sour cream. It might not be available in Belize. Now it is much better. Nine years ago, there were no groceries around. A stick of celery did not exist. You have to be able to adapt, not freak. You also have to be willing to accept the Belizean way of doing things, even if it is not your way. How they do it in the States is not the only right way. Adapt and respect their way of doing things. Otherwise you can't be happy.

"Finally, for Americans, nobody is impressed that you are American. That is a very hard awakening for every American. You need a work permit, or you need to get residency to work. Americans are so offended that they have to go through this. Belize is a foreign country. It is their right to choose and accept US citizens or not, just like the US does with Belizeans. Just because we're American doesn't mean we're special. Before you move to another country, think about attitude. If you can then that is great. If you can't you are better off maybe in the US."

PART I, CHAPTER 7: WORLD TRAVELER'S BAREFOOT BOOKS

Robert Henley and his partner Karen Boudreaux own San Pedro's first bookstore. Robert is British and grew up in England, south of London and then in the West Country. He has traveled, lived and worked in countries all over the world. He and Karen, who is American, settled in San Pedro in 2005.

EARLY WORKING LIFE

Robert's working life started in property management and land surveying. He then went to London and worked in marine insurance, then in the computer business.

Robert left England in 1972 and went to Africa, traveling to Kenya and further down to South Africa. In South Africa and Rhodesia he worked for a copper and nickel mining company.

Next, Robert moved to Canada. There he worked for another mining concern, in British Columbia. He left Canada after seven or eight months because of the cold. He hitchhiked down the West coast to the US and went to Mexico. Robert spent three months in Mexico and three months in Guatemala on Lake Atitlan.

Robert first came to Belize when it was a British colony and called British Honduras. "I hitchhiked to Belize with the

British Army from Tikal. I came here traveling in a fishing sailing boat from Belize City over the keys to San Pedro in 1974."

Ambergris Caye was little developed. San Pedro was very quiet, very clean with very few tourists. On his first trip Robert stayed two weeks.

The following years took Robert to numerous places, including Hawaii, Samoa, New Zealand, and South America. He lived in Arkansas for eight years, New Orleans for 20 years, and north Florida for a few more years. He taught and practiced Chinese and Japanese massage.

MOVING TO BELIZE WITH KAREN

Robert's partner Karen had lived and worked in Belize before, and likes it better than the United States. Robert had other reasons to move to Belize; Hurricane Ivan had destroyed the area in which he was living, he learned that he had thyroid cancer, and he decided to do something else.

"I met my partner Karen in autumn 2004 right after Hurricane Ivan. We came down in April 2005 for a couple of weeks, traveled around the country and then came back in November."

BOOK SHOP AND INTERNET TRAVEL BUSINESS

"We drove down from Florida. We just came in here into the bookshop, talked with Cindy, the previous owner, and discussed opening a used bookstore. Then Cindy said that her bookstore was for sale. We bought the inventory and fixtures and fittings. The building is leased and we added about 2,000 books in the following three months. We added a section of books about Central America and doubled the thriller and mystery sections. We also added a geography and maps section.

"The plan was to come down here and to start an Internet travel business. This means using the Internet to book travel to Belize, initially mostly from the US and Canada, in future also from Europe. For now it is difficult to get Europeans to Belize. Once the market opens up a little more, that will change."

Then they want to extend the business to Mexico, Guatemala, Honduras, Nicaragua, etc. They structured their corporations according to Belizean laws, creating several companies and an umbrella company called Lagniappe.

"We have this company and we have the travel business, Gaia Global Travel. It deals with travel within the country. We are a retail and wholesale travel agent with a website, and we joined all the organizations like Belize Tourist Board (BTB) and Belize Hotel Association. The wholesale part is that we are also selling to travel agents; you get a bigger commission if you are wholesale. For us this is an advantage and for the customer. We match customers to the kind of tours they want and the hotels they prefer."

Karen and Robert bought the bookshop because it fits with the travel business. It is local, a neutral business. They are not competing with any locals. They had never run a bookstore before.

"The two businesses work together well. I need to be here until we have enough business, then I will employ a manager and go on the road with the laptop and maintain the business that way."

HOW TO MAKE BUSINESS

"There are several interesting businesses on the island. Someone here in San Pedro sells real estate in Las Vegas. The possibilities increase with the computer and the Internet. There is a lady who does tarot reading down here, reading the cards. All kinds of business exist over the Internet. Some buy and sell stocks.

"I always had a large and expanding library and a large music library. I like books and music more than people. It is a little bit more reliable than people. If you want to know a lot about Belize you have to read lots of books by Belizeans, read all points of view. I bought a bookstore.

"I would definitely recommend making several preliminary trips. Be prepared that everything that has to do with the government takes six times as long or more than in the US. Patience is probably the most important thing."

If you are interested in Belize contact me at helga.peham@chello.at."

PART I, CHAPTER 8: STANDING ON HER OWN FEET

Nelly Brown is the restaurant manager of Lilly's Treasure Chest at Lilly's Hotel. Nelly originally came from El Salvador, where she was born and raised.

LIVING IN A LITTLE VILLAGE IN EL SALVADOR

"My dad died. This is a long story and it is the reason why I am here."

Nelly's family lived in a little village. Nelly's father was a vegetable farmer and her mother was a housewife.

In 1978, the war started in El Salvador. Nelly's family was very poor. The family stuck together and supported their father on the farm even as little children. Despite all the difficulties Nelly went to school, and graduated at the age of 18.

The family was still living in their village when the guerrilla war started in El Salvador. "It was so dangerous. The guerrillas were against the government. Not all people in the villages worked with guerrillas. The police came at night to our village and beat every single man." Every night they rang the bell so that all the men could run away to the mountains to hide.

"All our fathers and big brothers went away, into the bush, in the mountains, to hide from the police. One day everybody fell asleep and didn't hear the bell. The police knocked at the door and beat my dad very hard. They said that my father worked for the guerrilla movement. The next day his face was swollen. He decided to leave the village. We went to stay with his mother. All we took with us were some clothes and a few sheep, and there were the three of us, my sister, my brother and me. The war had spread all over the country but it seemed to be quiet there."

Nelly went to nursing school, where she learned to do everything from injections to blood tests to delivering babies. In Nelly's last year at school, her father became sick. He died in June 1989.

"At first, I thought that this was my end, too, and that I would have no more opportunities in life. But I finished school. My mother made me finish the year and I graduated. That was when I decided to go a different way to find work.

"El Salvador is a good country, but it is so hard to survive there." Nelly went away to find a job. "I used to work in the bakery in another village for one year, close to Honduras."

NELLY EMIGRATES

Then the idea formed in her mind, 'Why not move from here and go to the States?'

"I went to the United States with two girlfriends from La Union but we never got there; we didn't have the money. Instead we got to Belize."

Immigration had caught the young women in Cayo.

"If you have no one who is responsible for you it is difficult. We were crying. I was 20 years old. One of the girls said that her mother worked at Tabony in Belize City. Immigration called Tabony and found her. She went to Belmopan for us and paid the immigration fee for us. We then lived in Belize City with her for two months."

SETTLING IN BELIZE

Nelly found a job in a meat shop in Belize City. She worked there for six months. The wages were around Blz $85 or $90 a week, and rooms cost around $50 a week. A lady friend suggested they go to San Pedro.

"At first I said no, I don't know anybody there, and the girls sell themselves to Americans. But then I told my boss that I wanted to go and see my mother in El Salvador. He gave me Blz $50. That's how I came to San Pedro. The day we arrived, we rented a little room on Middle Street, just across from Ambergris Delight."

The lady returned to Belize City after a week.

"I had no money and I was so sad that I had to leave. Our landlady said, 'You have a future, you are young, you could do a great many things. Let the lady go back to Belize City. You can stay to clean the house and wash the dishes, and you will get a free room and free food.' I said, 'Fine, no problem.'"

CELI'S DELI

The landlady advised her to find a part-time job at Holiday Hotel, where her daughter was the manager. Nelly got a part-time job in the kitchen and washed dishes.

"My hands got really bad with washing dishes. When you need to survive you do everything, and it is only shame if you sell your body, work is not shame. Miss Celi, the owner, said, 'I could move you to the deli fast food.' I didn't know any English at that time. The customers were mainly Americans asking for sandwiches. You cannot learn English in one day." Nonetheless, Nelly joined the deli.

Hugo was the manager of the deli. "He told me to cook chicken for tacos. He took two chickens out of the freezer and I cut them. I had never cooked chicken. This was my first time and Hugo showed me how to do it, how much salt I need to put into it and everything else. I knew nothing about cooking. They helped me a lot, I learned how to cook a chicken, and a lot of other things, and they only had to show me once, I learn quickly."

One day, Hugo told her that it was time for her to go up to the window and take the orders. She had only studied English for one week.

"'I cannot do that,' I said, but Hugo said that if I am part of the team I need to take the orders. I went to the window, but I did not understand the language. They asked me about beef, a beef burger. This way, I was taught how to say it the right way. I learned a lot, I learned to cook and I learned a lot of English. Then Miss Celi said that I could have a full-time job if I wanted extra hours. I worked at the restaurant at night. I liked the job as a waitress, and it is a good job, if you like it. My job is very important, just as important as the job of a teacher."

They didn't pay her for holidays, though, and they did not pay her for extra hours. Then one day, they didn't even treat her as a person.

"One day there was a wedding on. In the deli we were five people working altogether. 'Nelly shall bring all the turkeys.' The owner told me to go and get all the turkeys. 'You have to go,' Miss Celi said. So I brought the first one. But I was so sad, because there were more people around to help, and I was asked to do it alone.

"I went down and I said to Hugo, 'I am leaving. No one else is helping to bring the turkeys.'"

MARRIAGE AND FAMILY LIFE

Nelly met her husband, Said, at Celi's place. Said used to buy a burrito every morning in the deli. At first Nelly did not like him because he was so quiet. Hugo, the manager, said Said thinks he owns the world.

Nelly started to talk to him. 'What would you like to eat?' she asked him. It was a burrito. Every morning he had his burrito.

After some time Nelly said to Hugo, 'You will see, this will be my next boyfriend!'

Hugo said to her, 'You are crazy! Look at you – you are an alien, he is a banker! He works at Belize Bank.'

Nelly said, 'Do you want to bet?'

'Are you crazy?'

'You will see!' Nelly replied with a smile.

"We started to go out. I invited him to my birthday party; it was my 25th birthday. We have been together ever since."

Nelly and Said lived together in San Pedro in Belize.

"When I got pregnant, it was time to get married. He said, 'Let's get married. Marry me because we will have a kid, and I want to have my kid on my lap.'

"If he treats me well I will stay forever, if not I will continue on my own way. You get married in church once in a lifetime only. I decided to get married only by civil law.

"For me being married does not mean receiving orders. I told him that we both bring money home and whatever we can do we do it together. We work for the same reason, have the same focus. Said thinks differently about it. Men think that if you have a wife she will do everything for you. That is the macho thing. We need more communication. We need to talk about what I like and what I do not like. I do not like his demands on me. For me, it is like a knock on the head."

RESTAURANT MANAGER

When Nelly left her job she was sad to leave all her friends. After a few months, she found a job in the Paradise Resort at the snack bar. She worked with an Italian lady.

"She was a nice person, I think she smoked weed, sometime she was nice, sometimes she screamed. She gave me the job, but when she had a hangover she treated me badly. So I started looking around and I found this job eight years ago in Lilly's Hotel and Restaurant."

At Lilly's, Nelly started as a helper to the cook in the kitchen. "I don't like the kitchen; it is too hot. Any space outside, and I am happy."

One day the waiter was drunk. They asked who could help as waiter that morning. Nelly had always been asking for this job, so they called Nelly.

"We were always two to three people working in the restaurant. People come and go. You cannot work in this place and drink – it is not possible. We used to hire men, but the last one drank, so I got the job. That was in 1998. My baby was 10 months old."

A HAPPY LIFE

"Myself, I have a happy life. Nothing is bothering me." Nelly likes to help others. "When you help somebody it is not because you want to receive, but only because you want to give. My goal is to be happy in my life. You have to be who you are, and not to try to be somebody else. I just think I am happy the way I am because I always have dreams and plans. If I want something I can do it. Everything is possible if you have a focus about what you want in your life. I have a lot of friends. My husband is angry at me about that. I just want to see people and talk to them. When you are happy with who you are you spread happiness around. Love yourself so that you can love somebody else. When you feel insecurity everything will go wrong. You have to love yourself."

VISITING HOME

Nelly has strong family ties. She still communicates with her family regularly, frequently calling home on the phone.

"I used to help them when I came over here. Now, I don't have to send them money anymore. I helped my brother and sister to start up a chicken farm. They have a better life now. I still help my mother. She is 61 and she is always sick. I help her to get medicine and anything else she needs. My elder brother lives here. He built his own house in the same yard on our land. He is married and has a wife and three kids."

In 2005, Nelly made a surprise birthday visit to her mother.

"I reached the airport in El Salvador and bought a cake before going home in a taxi. My sister had prepared everything for the birthday party. That morning, my mother went to town to buy food. So I hid. My mother was so happy when she came home and saw me and my daughter. 'Nelly why did you do that? You are so good to me.' I have plans to go this year, but you never know."

NELLY'S GOALS IN LIFE

Nelly has dreams and plans for the future.

"I like to save my money. I have my own money that I can depend on. I have my own savings, if something should happen. I would like to have a second baby soon. My first child is eight years old already. She asked me, 'Mummy, do you think I am old enough to have another brother or sister? I really want it. Are there no more babies in the company? My aunty took the last one.' I am ready. You need a significant amount of money for a baby. I want to have my second baby before I am 40."

Nelly learned to use a computer at Introcom. She exchanges emails with friends in other countries and wants to learn how to shop through the Internet. "I never finish putting new goals into my head. My life was so hard, but I am happy with myself now. I am not happy in my marriage, but it is not bothering me. As long as people around are friendly to me this doesn't bother me."

Nelly likes San Pedro a great deal.

"I think it is the freedom of this place that I fell in love with when I came over. Now this is the house where I live, and I have everything here that I need. Most people treat you well. People give you a big smile and show you that you are welcome. It is small, but there are a lot of things you can do here.

"You have to look right in front of you and not behind you, and nothing will be impossible!"

PART I, CHAPTER 9:
A LAWYER PLAYS MINI-GOLF AND SELLS SOFT ICE

Allan Pentkovsky was born in New York City and raised on Long Island. His life changed completely after he moved to Belize. He was a lawyer. His wife, Patty, is a former flight attendant. They now own a mini-golf course and two ice cream stores in San Pedro.

A LAWYER IN HIS OWN PRIVATE FIRM

After law school, Allan worked for a small law firm in central New York for three years. In 1979, he established his own firm in which he worked for 26 years. For the last 14 of these years, Allan was a bankruptcy trustee. His practice was in the field of corporate law. The firm handled the liquidation of many businesses and sold a wide variety of items – horses, farms, restaurants – and became involved in many of these businesses. "I got very much into business."

WINNING AN AUCTION

"Patty and I traveled frequently. She was a flight attendant for US Air after we got married, so we traveled a lot. Whenever we traveled we thought of moving to an island, opening a little shop. Our first trip to Belize was about 1999. The Public Broadcasting System PBS in our area had travel auctions on

TV. One day an auction came up for the Sunbreeze Hotel in San Pedro, Belize. At that time, we did not even know where Belize was. We bid on the trip and won it, and so we had a week in Belize at the Sunbreeze Hotel. That was our first trip to San Pedro."

FIRST IMPRESSION OF SAN PEDRO

"When we first landed, Patty was ready to take the next flight out. But there were no other flights; we had taken the last flight of the day. We went across the street, then into the hotel. It was on the beach and had an ocean view. She immediately changed her mind. We stayed a week and had a wonderful time. We raved about it to our friends and we came down the next year with three other couples."

The little party of eight friends returned the following year. They came across condominiums under construction and ended up buying one, far south on the island of Ambergris Caye. "We made an offer and bought one. Two couples bought another one together. That was in 2000. Once we had the condo we came down more often. We were divers. We all got certified by YMCA. We enjoyed snorkeling, diving, all the tourist things you can think of for a week, ten days at a time."

MOVING TO A NEW LIFE

Allan and Patty got to know the island and met people.

"We made the final decision to move to Belize in March 2003 while on a vacation. We wanted a business that the kids would enjoy, and the land became available at that time. We decided to try a radical change. I sold my law practice, our home, our cars and our furniture. We moved in September 2003."

The couple first opened a mini-golf course with an ice cream shop. Their second project was a clothing shop, but they closed in August 2005 because costs outweighed profits. With import duties in addition to sales tax and shipping, clothes from the US were too expensive to sell.

But Allan and Patty learned what sells on the island, so they opened the second soft ice cream store right in town in November 2004. Mini-golf plus ice cream has proven successful.

"It was a big, big change for us. We missed our friends. There are not a lot of cultural things down here, no movies, no theater. On the other hand the pace is very slow and there is a very low stress level."

It took them just a little over four months to complete the move and get their business licenses.

"It's always good to know some local people to help you get settled."

"We are fortunate to have some very good workers. We contribute to the economy and we help. Mini-golf was open for two weeks for free during Christmas vacation so that the kids could play.

"For anybody who is moving down here there is a long learning curve, and with business it's even longer. People have to educate themselves to import goods, and get a reliable customs broker."

The couple adopted an island dog.

"Actually Molly adopted us, and we have a parrot named Buddy. She learned to bark. We have four dogs that live on our property and five who bark."

BUSINESS OPPORTUNITIES ARE EVERYWHERE

"Whatever is lacking here is a good opportunity. There are no international newspapers here. Bringing them in and selling them in a little shop would be a good chance. We have very little manufacturing in Belize. This could be made available on a larger scale. For example, we could make shoes, clothing, sneakers, etc. There are many good laborers here in Belize. Somebody opening manufacturing plants would do well here."

ALLAN'S ADVICE

"The island of San Pedro is a beautiful place. The government needs to think about its infrastructure. This is a gorgeous place for people coming on vacation.

"My advice would be to come down all different times of the year, not only during the high season, and spend an extended time here. You have to come down knowing that you are a guest in another country. You have to accept their ways."

PART I, CHAPTER 10:
A DREAM LIFE

Ana Rivera Lausen was born in El Salvador. In her childhood, she also lived in Belgium and Spain, but lived in Los Angeles for most of her life and considers it her home. She worked in Boston as a marketing manager. Ana has traveled widely through her work. "I went all over the world. I love to travel and to scuba dive."

A LONG HOLIDAY

Ana made a spontaneous decision to dive at the beautiful reef in Belize. She decided on Friday and came on Sunday.

Ana stayed at a Bed and Breakfast in San Pedro. After a week she asked the owner if she had a room for another week, then another week, then another week. At the end of one month the landlady asked Ana if she could stay for another month because she wanted to go to Australia to visit her sister. That was in May 2000. Ana said "Sure." She stayed for no pay – just room and board – to run the business for a month. It was Ana's first experience in the hotel business. It was a six-room hotel, Changes in Latitude. The Canadian landlady became one of Ana's closest friends.

"I was lucky enough to meet several other strong women who were living here in San Pedro. Everybody ended up here for different reasons. A special bond of sisterhood helped me not to feel lonely. We did a lot of scuba diving, a lot of reading. Previously I had friends but I did not have so much time to dedicate to my friends. I had that time here."

SEPTEMBER 2000

Ana had quit her job before coming to Belize and had planned to go back to graduate school to get a PhD. Ana's friend, the hotel owner, decided in September 2000 to go to Canada for a vacation, and to let Ana take care of the hotel and her home upstairs again. However, Hurricane Keith destroyed the house. "After the hurricane I realized how much I liked the island. Everyone helped to build up again. Then I decided to stay."

Keith also taught Ana that she was stronger than she thought she was. "My girlfriend and I prepared for the hurricane. We had enough food and water for ourselves for two days, but when it was over we were fifteen people. We shared food and water with fifteen people and got to know them very well. You realize how little you really need, just a little food and water, clean water."

A BLOND DANISH MODEL

"Three weeks later I met this wonderful, most gorgeous man I have ever seen. He was a model in Denmark. He talked to me one night and we talked all night, then we started dating. Eric, who was visiting his sister for three weeks in San Pedro, went back home to pack his things, then returned to live with me. We met in November 2000. We were married two years later, in 2003."

"My mother and my friends were concerned about me, if I had completely lost my mind, what was I doing on this island, in this country, but we went to work with Eric's family, at his sister's business at TMM Ltd. Yacht Charters.

"I gave birth to our daughter Alexandra in December, and her brother Nicolas was born in July 2005. We do get island fever and have to leave the country every few months. After a week we want to come back. Once a year we go to Europe and several times a year to the USA."

Eric's mother is from the Bay Islands, from Roratan. Eric was born in Honduras and speaks fluent Spanish. He was 8 years old when his parents divorced. His Honduran mother moved to Denmark with the kids while his Danish father stayed in Honduras, but Eric never forgot the island or the Caribbean life. He feels more at home in the Caribbean than in Denmark.

"Our fathers are in El Salvador, our mothers are very much alike, the same interests, really funny, our mothers are best friends, really good friends. When I first met Eric's mother it was like meeting my mother. When he met mine, he thought that she was like his mother."

Ana's and Eric's children have four citizenships: American and Salvadorian from Ana, and Danish and Honduran from Eric.

LIVING IN BELIZE

"I had been all over the world. Belize was the one place where I immediately felt at home. The people were so warm and so welcoming. They were so kind to me. There were even local families who opened their homes for me, made me feel at home, made me feel I belonged here. I'd heard from other people that the first year was difficult until people accepted you, but I didn't experience that. People here are so giving. Nobody asks you what you do for a living, what kind of car you drive. Here maybe it took three months until they asked me what I do. Here it is to find out who I am, not what I do.

"There is not a big difference in social classes. Everybody talks to everybody. The richest person on the island talks to the person selling cashews on the island. I saw Dixie [Bowen, the owner of the private primary school 'Island Academy' and the wife of Barry Bowen, the richest man in Belize] talking on the beach to the guy selling cashews, no shoes."

A CHANGE IN PERSPECTIVE

"I have realized I am now repelled by materialism. I am now horrified when I think how materialistic my life used to be, always buying more. I thought I needed to have 80 pairs of shoes. You don't need stuff. Here we don't have stuff, we have what we need to be comfortable, not more. I now work to live, I work but I make time, quality time for my family and my friends; my life is my family and my friends although I love my job.

"I have learned a lot from people who before I wouldn't have given a second look to. I have learned from the most unlikely characters, that before I would not have given the time of day. We are all important, we are all connected, the whole world, everything happens for a reason, sort of. According to quantum theory, we are all connected.

"I have very good karma here. I always had very good luck in my life. Now I see more and feel more how lucky I am."

Ana once called information to ask for a telephone number. "The operator said, 'Hello, Ana,' because she recognized my voice. Where else in the world can that happen?"

WHY BELIZE?

Ana started to ponder when asked what made her stay on Ambergris Caye. "I don't know. It was after the hurricane and I saw the island at its worst. It couldn't be worse, but still I wanted to be part of it. I worked to rebuild the island, I felt part of the community. There can be days when I wonder why I'm here, what I'm doing here, but these days are few. I just look out to the sea, and then I know why I'm here."

There are no special plans for the future. They may have to leave for the kids' schooling when they are much older, but they don't know where to go, not even which country.

"I'm afraid I wouldn't fit in the Danish lifestyle. I'm also afraid that in ten years we will not be able to live on this island. Housing is a real problem for the locals, the cost of living. It's hard for the locals to buy something now while it is still semi-affordable, and I consider myself a local now. Real estate prices have doubled, in some places tripled, in the last six years. I am sure it is cheaper now than it will be five years from now."

WORKING FOR TMM AND VICTORIA HOUSE

Ana used to work at TMM, a worldwide sailboat charter company with a base in San Pedro, with her husband. Now she works at Victoria House, a luxury boutique hotel, as the Assistant Manager.

At TMM the couple managed a fleet of 20 catamarans. This was their first experience with the tourism industry. They were supplying a first world service, and the island's resources presented them with unique challenges. People came to TMM expecting a luxury vacation, and Ana and her husband had to do many things behind the scenes to ensure everything was perfect when the customers arrived.

Victoria House runs the highest occupancy rates in the country year round. "Our restaurant is one of the best in the country as well."

Victoria House also offers unique wedding packages.

"We have lovely weddings down here, several weddings a months. I coordinate the wedding details with the brides. We do beautiful weddings, only for Victoria House Guests who stay with us. We do them in many places: around the pool deck, on the beach. We have a bridal suite in this house. The basic package costs US $1000, up to tens of thousands."

Information helga.peham@chello.at.

ANA'S ADVICE

"Come down here with an open mind, without preconceptions, accept this country and its people for what they are. People get frustrated when they expect things to work the same way they do at home, wherever their home may be. Don't expect things to be the same. Why not accept? It's their culture. Feel lucky that Belize opened up its arms to you – we should just appreciate that instead of getting frustrated. Sit and write down the little things that can only happen in Belize, and look back on the stories and laugh. Take it with a sense of humor and a lot of patience and with an open mind. Very often funny things happen to us that make us shake our heads."

PART I: CHAPTER 11:
NEWS FROM SAN PEDRO

Tamara Sniffin lived in Venezuela and Costa Rica, traveling all over South and Central America with her parents, who were in the Peace Corps. She has degrees in Social Work and Fine Arts. She worked in the music business and became Director of the Wyoming Children's Museum and Nature Center, later moving into freelance fundraising and grant writing.

Ron has a degree in broadcast journalism and worked in Wyoming as a reporter and videographer. He was the Communications and PR Director for the National Education Association until 2001. Later he and Tamara moved to San Pedro with their son and bought the San Pedro Sun newspaper.

(Learn more about the newspaper by contacting helga.peham@chello.at)

GILLIGAN'S ISLAND

"My family has been traveling to Belize for 30 years," Tamara explains. "My parents bought a house here 27 years ago, and we traveled here almost every year for vacations, so in a way I grew up here."

Her parents' house was right behind Banana Beach Hotel. There was no electricity at that time, so they used generators. In town there was current for a couple of hours per day. The family got drinking water from a well, and they also collected rainwater. They used kerosene light at night. A little path led through the jungle to town, or they walked along the sandy beach.

"It was all very, very rustic and really reminded me of Gilligan's Island – quiet, pristine, fabulous. San Pedro was so secluded and simple, life was very simple and quiet, really a good place to relax, to just enjoy the water and the sun."

Every time the family returned, there were new resorts, businesses, and roads – growth was very noticeable from year to year.

FANTASIZING

"We had always fantasized about living here. We subscribed to the San Pedro Sun. Then it was up for sale and we went into a business we knew something about which would allow us to work here."

They had traveled to Belize many times. When the opportunity came they looked at houses, talked to people, and learned what it would take to live here. Tamara's parents had a wealth of information because they had lived on Ambergris Caye, south of San Pedro, for a long time.

SAN PEDRO SUN

"We saw the opportunity and took it. Both my husband and I love to travel. We didn't want only to talk but to make it happen. It was a big decision to buy the paper and move here. My family was very supportive, knowing Belize. My husband's family thought we were crazy. Many come to visit us.

"We realized our dream. San Pedro Sun was a modest publication and we saw a lot of potential for growth. We were pretty motivated and made some changes. This was an interesting way to get to know the community better. It was different to being a visitor. Working here is different. You are not on a vacation. It takes the fun out of it a little bit. But it's great to get to know people quickly, to become a part of the community really fast.

"We are definitely community oriented. We include a visitor section, features on places to see, investments, cultural information, travel information, street treats, a different food every week... We present different fruits, street vendor food. Thus we familiarize the visitors with things that make this area unique.

"The San Pedro Sun reports on community news, on everything that happens in the community. We report as unbiased as possible. Everybody has the opportunity to have their voices heard.

"The office receives emails, phone calls and faxes. We have a staff of four. Three of them write. They hear the gossip and the rumors and decide what to follow up. Their job is to hear. We have good relationships with public officials. The police give us a report every week, and other officials also keep us informed."

They publish a section in the 'Visitor Guide' called Lizard Tales, which features stories of superstition and folklore. "There is no shortage of stories." They write the stories in Creole, Spanish and English.

SAN PEDRO NOW AND IN FUTURE

"San Pedro is growing so fast; the town councilors are working very hard to manage the growth. I hope it can be done responsibly. We seem not to have the infrastructure to accommodate this kind of growth. I hope Ambergris Caye keeps its uniqueness. Some people are concerned it may become a second Cancun. People who live here want to keep the culture as it is.

"I just think that Belize in itself is one of the most culturally rich countries, with its Mayan, Creole and Garifuna influences. The Central American countries have a great mix of cultures. Everyone is hanging onto their own traditions, instilling those cultures into their families and sharing them. Garifuna celebrations happen here every year. Historically it is a fascinating area."

FESTIVITIES

The newspaper staff tries to emphasize vacation planning whenever they cover special events.

"The last weekend in June is the 'lobster fest' in Placentia, and it is in Caye Caulker in the first weekend of July. They have a Miss Lobster Fest. You get any kind of lobster food you can think of. It is a food and fun weekend. People come from the US every year, especially to the event in Caye Caulker."

Another is the Costa Maya Festival in San Pedro during the first weekend of August, which features food vendors from all the Central American countries.

"These festivals are a great way to immerse yourself in the culture and enjoy what we have to offer here.

"People celebrate Garifuna Settlement Day all over the country. Dressed in traditional Garifuna dress, they come from the water and re-enact the historical battle, then perform a ritual with traditional music, food and dance. They have a Miss Garifuna Pageant. She must speak in Creole. It is a fabulous experience."

PUBLISHING THE NEWSPAPER

"We have seen incredible progress in technology. When we bought the paper, we pasted up the pages, put them into a big cardboard envelope, and called a cab to pick it up and take it to the airport. A courier picked it up at the airport in Belize City and took it to the printers. We did the paper in parts, by page segments." Each stage was another opportunity for something to go wrong. "Now we can email the printer our documents. They do everything. This makes life so much easier."

However, sometimes the Internet goes down.

"Once there was no newspaper in the country, because there was no paper in Belize. My husband traveled to Mexico and Guatemala to acquire newsprint to print the newspaper. That was September 2005. Nobody had newsprint. There were a couple of weeks where we did not know if we could put out the newspaper."

EDUCATION AND INFRASTRUCTURE

If people do not have the money, they do not go to school. Local government supports education as much as they can, but they do not get much money from the national government.

"San Pedro contributes the vast majority of income taxes but gets only a small return. This is very hard on the island; thus our streets look the way they do. There is the complaint that not enough qualified people are available. This is a young country; it is only 25 years old. You have to keep that in mind when facing problems.

"When Belize was British Honduras, British soldiers often vacationed in San Pedro. There were British soldiers in the bars, and some were passed out on the beaches. The military was very visible. In 1981, when the country became independent, you could see that the people of Belize were really happy to make this country their own. But progress came to a stop, to a screeching halt. Of course all of our funding has changed since then. Priorities have changed as well. The original infrastructure was certainly British, in education and other areas. But both then and now, it all goes back to politics, where the priorities are set by the politicians.

"One of my favorite stories is from the early '80s. I was staying down here for several months with a girlfriend. I had a friend who flew in from Corozal, but we couldn't meet her because we didn't know when she would come. We were sitting on the porch when suddenly someone cried from down there. We looked out the window and saw my friend. We hired a man with a wheelbarrow, put her luggage in it and went down to Banana Beach.

"First you had to come to the island by boat. Later, a connecting flight from Belize City was established. One of the first times we flew in it was at the end of the day, getting dark. There were no lights on the runway. They rounded up the vehicles and turned on their headlights so that the plane could see where to land – a truck and a few golf carts, just enough light to land. This was the beginning of Tropic Air."

LOBSTER AND FISH

"I remember when the fishing co-op was full of fish and lobster. Lobster and seafood is not as available as it used to be. They are trying to install fishing seasons for lobster and conch. Reef management and conservation education are necessary. The Peace Corps are involved in marine biology and education. They assess and keep records of the reef's health, working to maintain it. A lot of education is necessary. The tour guides know that the reef is essential to their lives. Reef management is important. It's readily evident that the amount of fish around has decreased tremendously. They don't open lobster season until June 15th. There was a drastic reduction on the first day this year.

"When I talked to director Milo Paz, he said it was probably overfishing. There are people who harvest lobster. Marine biologists saw kids bring in small conchs – a boat full of baby conchs. We need to educate and communicate; this is so important. They need the foresight to know that what they are doing today will affect their lives. They will feel the repercussions in their future for actions taken today.

"There is much medical assistance on the island, including dentists and other specialists. There is even a program to take people with eye problems to Cuba to get help there. The medical volunteer help here is great. The island enjoys a great deal of outside support. Education is an area where help is needed, and not just through books and supplies."

TAMARA'S ADVICE

For people who want to move to Belize, Tamara recommends not selling everything and burning every bridge.

"We kept our house, rented it out. You need a safety line. Some people sell their home, everything. Sometimes they find out they don't like to live here, then it may be difficult to go back. Live temporarily down here for several months before moving. A vacation is too short. Especially, living on an island is different. A whole handful of problems, like getting things here, plus there are other changes when you are in another country."

What about business opportunities in San Pedro, Belize? There are still, obviously, good opportunities. Expatriates from the US and other countries continue to move here.

"There seems to be a need for products those people are used to, like espresso and cheesecake. If someone did this, it would be great. It is nice to have these comforts from home. A used bookstore was an excellent idea that was just realized. No more realtors, and no business that directly competes with locals. No businesses are needed that are covered by locals."

Tamara also advises, "Be selective about who you use when you want to do business.

Seaview

Doggy and Three Puppies

Fisherman after Work

Palms

Kind Old Man

A Belizean Woman

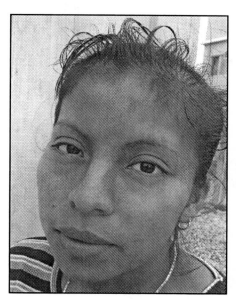

Young Belizean

TMM Fleet

PART II: BELIZEAN FRIENDS

PART II, CHAPTER 12:
CARING FOR DOGS AND CATS

THE CAPITAL OF BELIZE

David Alvin Warren Bellini was born in Belmopan, the capital of Belize, on April 20, 1987. His father is an Italian American, his mother is from Guatemala, and David is Belizean. When he was 4 or 5, his family moved to Miami. Two years later, they returned because they missed Belize.

CHILDHOOD IN THE US AND BELIZE

"We grew up here in Belmopan when we came back from the States. We all started school here in Belmopan — my two younger sisters and I. They are still living in Belmopan."

David attended Belmopan Comprehensive High School. Two weeks before he graduated fourth form, he quit. His mother wanted him to stay at home during the day, but David wanted to go out and work, so she kicked him out of the home. His father had left three years earlier.

At school he was first in class in Business and learned the principles of commerce and office procedures. Later he switched to science, becoming interested first in physical therapy and then in grooming animals.

PAMPERED PAWS

When David first came to San Pedro, he worked at Belizean Shores as a chef. "It was fun to work there despite the long hours, from 8 in the morning until 3 the next morning. I worked for a month and half as a chef.

"The bartender knew I liked animals. So when the young guy at the pet shop quit, he told me. I decided I would go and check this out. I called and came the next day for an interview. I got the job right away. I came to work in Pampered Paws from 8 to 5 and then worked 6 to 3 in the resort."

David also sang on Wednesdays, Fridays and Saturdays at Fido's. This was a lot of work. David did this for four months before quitting Belizean Shores to concentrate on his work at Pampered Paws.

"About a month later I came to work full time from 7 am to 7 pm at Pampered Paws. When there is much to do I even work through lunch. It has always been my habit."

David takes care of about 48 dogs per week.

"Most people don't make appointments when they bring their dogs in. I cut their nails, sometimes paint their nails, clean their ears and brush their teeth. We shave their faces, poodle style or what the client wants. Sometimes the dogs come with fleas and ticks and with eggs. Sometimes I spend five hours taking them all off. Sometimes they are really covered in them. We give them flea baths, medicated baths, depending on their skin. Or we give just regular baths."

Other services are outdoor spraying. When the clients'

yards are full of fleas, ticks and other insects they go and spray the yards. They order and sell necessary food and supplies for the animals: clothes, treats – whatever the dog and cat owners want.

Pampered Paws also boards dogs overnight. If the customers don't want to bring their dogs to the pet shop, David will stay with the dogs in their condos and walk them as needed.

PARTY TIME IN MINI-MANHATTAN

"This is quite a demanding and interesting job. After work I just go skating. I go roller-skating on the cobblestone. I know how to skate almost anywhere." San Pedro's Middle Street and parts of Front Street are newly covered with cobblestone. "That is my passion. I used to play hockey in Belmopan. My mother didn't like me playing hockey. That is why I got kicked out from home."

David doesn't feel lonely in San Pedro.

"I just made friends recently, this place where you party every night; they take me to places every day. Every Thursday to Sunday there is a party every night. Thursday, they call it Thirsty Thursday. We go in golf carts everywhere and buy drinks. We go to Fido's, Jaguars, Big Daddy's, Barefoot Iguanas, BC's 1 and BC's 2 on the beach, Cannibal's, Tackle Box; there are so many places. Friday and Saturday night are for dancing. Sunday, we sleep almost all day, then wake up and go partying again.

"Sometimes they go on Monday as well. Some people are paid on Monday and they want to go partying and drinking right away. Our group are me, my roommate Eric, and all the rest are girls. At the club I dance for a band sometimes. I do choreography. I dance at Jaguars and Barefoot Iguanas.

"Before I came here my girlfriend went away with a guy I called my best friend. I have to get over a lot of things. I recently got out of another relationship. Now I am just back on the road again partying.

"If I go to the US I will study and graduate in grooming. I want to take the high school equivalency test, and then I will go to a six-week grooming classes and get a diploma. I will probably do that in a year or a year and half. I just need to save enough money to go to the US and come back." (helga.peham@chello.at)

PART II, CHAPTER 13:
THE MAYOR IS A LADY

"I am the fourth female elected on the island," explains Mayor Elsa Paz. "I was born in 1963, here in San Pedro."

A FISHING VILLAGE

"In my childhood and youth in the 1960s and '70s, San Pedro was a small fishing village. It was totally different; very sandy streets, two main streets, front and back street, Barrier Drive and Pescador Drive. Now we also have Middle Street, the third and fourth streets. In my childhood there were lots of mangroves and coconuts.

"The village expanded from where we have the Belize Bank now all the way to the primary school. Pescador Drive was from where we have the electricity plant, BEL, all the way to where they put the primary school. There were no vehicles when I was a little girl of 9 or 10. After a while we had two or three taxis, land rovers.

"We used to have a fishing cooperative. The only industry on the island was the fishing industry. San Pedro had one of the most progressive cooperatives in the country. The cooperative had some cargo boats – Elsa P., Emma V. With that boat we went to Belize City; it took almost six hours to get there. There would be a big sailboat with a motor. Until I was about 12, 13, there were no airplanes here on the island.

"We used to have a lot of sailboats in the cooperative. Fishermen went on trips 10, 14 days at a time to get lobster, conch fish. That used to be the bread and butter of the island. Most of my uncles and even my oldest brother were fishermen. My father was one of the first non-fishermen. He used to sew and he had a barbershop. All my other uncles went with my grandfather fishing.

"My father didn't like fishing, cutting hair, or sewing. He had one of the first bars and entertainment places on the island. He used to have a saloon and a bar and a dance hall. The entertainment place was big; it's where Big Daddy's is today. He also used to own one of the biggest stores on the island, one of the biggest grocery stores, a general store, where he sold groceries, equipment for fishing and hardware. He also was the one who opened the first tortilla factory. Today my aunt owns El Patio.

"My mother was a housewife. We were eight children. My father passed away in 1972. My mother raised all of us."

OWNING AND RUNNING A GIFT AND ART SHOP

"We used to have one primary school, the Roman Catholic school. All children attended this school, for nine years. After that I went to high school for four years.

"After high school I worked for five years as a receptionist in one of the hotels, the Barrier Reef Hotel. For two years I worked at San Pedro Foods, one of the first supermarkets that we had.

"From there I started my own business. I opened my own place, one of the first gift shops on the island, with a lot of artwork. I was an artist before I was mayor. I made hand-painted T-shirts, wooden carvings and paintings, and sold everything in my gift shop. I had it for 17 years. I took courses in Mexico on arts. That is my profession, art. I learned how to do art in Merida with wood, aluminum, wood carving. Then art like painting. I also learned to paint on ceramic, ceramic arts.

"I went to an art school in Merida for six months to learn different ceramics, carving, and more basic things. My cousin, she is good, went to study for five years, and became a professional artist.

"Most of my customers were tourists. Tourism started in the 1970s. Holiday Hotel was one of the first hotels on the island. John Greif first came in with his airplane in the 1970s."

CALLED INTO POLITICS

"I first got into politics in 1988. I ran for Councilor and lost. At that time people elected seven councilors who would then elect the mayor. 90% of the island supported the other party, PUP.

"One of my cousins and one of my very good friends ran with me. We ran just to have an election, and we lost. During our campaign and rally, the leader of the UDP party, Manuel Esquivel, prime minister for two terms, won. In 1994 UDP won again."

MAYOR OF SAN PEDRO

"2003 was when I was elected as major.

"I have a lot of plans for San Pedro. One of the reasons to run for another term is to finish my projects that are still pending, like the renovation of Central Park. The Council decided this is a family park. This used to be the basketball court that is now at Boca del Rio. We are working on Central Park, the Plaza. Families will enjoy this, go there with their children. There are three different playgrounds there. Central Park will also be used for public events. The Council wanted something more attractive for the tourists and a tourist information board there.

"We do not have so many plans for the other side of the island. We have already completed the bridge. There was a big demand for a bridge, because the hand-driven ferry created a lot of problems for construction workers in the north. They often had to wait for hours to cross with the ferry, before they worked each day and again after work each day. Most of the construction work is on the northern island. Only pedestrians and golf carts may cross the bridge.

"In the north there is an airstrip that isn't used. There's a lot of bush there. We have the bridge; no vehicles are allowed on the other side yet because there is no good road. We know that in the long term it will happen; a road will be built there, eventually there will be traffic. We have to be realistic that the largest part of the island of Ambergris Caye is in the north. The whole island measures 25 miles in length from the bridge: 18 miles north, 7 miles south. We will need to develop in the north. In the south there are no government properties, there is only private property.

"In the north there is a lot of property by the Basil Jones area, 9-10 miles. This is under Belmopan. All is at the back of the island. Foreign investors own most of the beachfront. Eventually the north side will be developed. The land will be distributed to other generations. The Council will get the government to subdivide land. We cannot stop development, but we can control it.

"I will encourage investors to open businesses that will be beneficial to the youth. We don't have recreational centers. One of our goals is to build a small civic center. Investors would be welcome to open a business to entertain the youths.

"What we do against drugs: I created a safety committee that works hand in hand with the council to eliminate crime. The safety committee is working very closely with the council. We try to work together with this committee, the police department, hand-in-hand with the Minister for Home Affairs. Police Department, Immigration, Customs, Labor, and Fire are all under Minister for Home Affairs. We do as a council try to work together with them, assist them; they are working for our community. The best thing is to work together. We have meetings with all of these departments, every three months, sometimes even more often, but drugs are not under our jurisdiction, it is under the Ministry's."

HOW FOREIGNERS CAN MAKE SAN PEDRO THEIR HOME

"I believe that our island is a very peaceful country. Compared to other countries in Central America, our island is the most peaceful place to be as a foreigner. I believe they will become part of our island. Learn to live with our culture, traditions; they will go along with whatever tradition, culture is offered. This is a very united community; we try to help each other. Our industry is tourism. If we don't have tourists, we don't have a living. Our community spirit will help build the project. We are number one fundraisers for the whole country in Belize. If somebody is sick, we unite, we try to help. We are very united. That's what makes San Pedro strong.

"I am here for the community. I want the best for the island. That is why I ran for mayor. Whatever project I can accomplish, I am here to give San Pedro a better image. I want to keep San Pedro as number one. My goal is to accomplish many projects beneficial to the citizens of the island. For our children to have a good level of education, youth off the streets, getting them involved more in sports, focusing much more on the youths."

PART II, CHAPTER 14: ONE OF THE OLDEST PEOPLE IN SAN PEDRO

Her story is told partly by her son.

EARLY YEARS IN ORANGE WALK

Mrs. Regina Gonzalez was born on September 7, 1907, in San Roman, Rio Ondo, Orange Walk. It was a village of about 300-400 people at that time. She lived in San Roman until she was 17. She went to school until she was 14 years old.

When they met her husband used to go by boat to San Roman. He went into the bush, drained the juice from the sapodilla, and sold it to Chicklet's Chewing Gum. The contractor, who was from Orange Walk, paid him according to how good it was.

GETTING MARRIED IN SAN PEDRO

Regina's husband brought her to San Pedro. She was 17 years old. She had never visited San Pedro and knew little about it. Her wedding dress was sewn in San Pedro using her measurements. She didn't know how much money her new husband earned, either.

COCONUT PLANTATIONS

He owned land with coconuts in San Pedro, so they worked there. San Pedro was smaller then, extending only from the primary school to what is now San Pedrano and Lilly's. In the back, the mangroves were so thick they had to cut through them to get to the lagoon.

First they lived in town, then they built a house four miles north on the coconut farm and lived there. A lot of families moved there. Everyone worked chopping and clearing the land. Mrs. Gonzalez and her husband had to clear two pieces of land before they could plant the coconuts. She helped him before she had children. They eventually had seven children, all boys. Their eighth child, the first-born, died young.

Big boats took the coconuts to Belize City to sell. Belize City then exported them to the US. At that time 1000 coconuts sold for Blz $100. They picked and sold 8000 coconuts a month, earning $800, which was an excellent amount of money back then. Then the price dropped from $100 to $8 in 1931. Many places, such as Honduras, had too many coconuts, which is why the price was so low.

HURRICANE 1931 AND THE HONDURAN CHALLENGE

In 1931 a hurricane came, destroying Belize and San Pedro. The thatch of their house was blown away. The coconuts were damaged, so they had to replant, but it was difficult because earnings were so low.

SECUNDINO RECALLS

Secundino, one of Mrs. Gonzalez's sons, was born in 1941. There were two schools in the area. Secundino recalls that the teaching at the school he attended was very good.

"In my days we had the British Army in Belize City. They only taught English at school, no Spanish. The teachers were from Belize City and Dangriga and they only spoke English. They were Belizean but they came from the Caribbean Sea. They came to the island to teach, and they were smart. The British Governor financed the schools and gave books to the children. In those times the children were taught more discipline, children paid attention in school, the teachers were strict. Not many children and one teacher, not many books. They used to teach mathematics, spelling, and reading good English."

He attended primary school until he was 13 or 14 years old. When he was 14, he began working.

FROM THE COCONUT FARM TO SAN PEDRO TOWN

Everything was quiet in Belize during World War II. The British army came only when the war was over.

In 1942, another hurricane destroyed the island, and the coconuts again, so they decided to move to San Pedro town. But it was hard to get money, so the family went to Mexico to work there, as did many people from the island. The coconuts were good there. They chopped for a few months, then went to other places to do other work, such as cutting mahogany south of Belize City.

"When we went away after the storm," Secundino recalls, "Times were very hard. My father died when he was 56. Then things were worse for our family. Children of all ages had to work; it was hard, nothing to eat. Sometimes the neighbors gave food. Often only flour, only corn tortillas, nothing else. We still had the land, but the coconuts were very cheap, and we only got 100-150 coconuts every week or two, five cents for the big ones, three cents for the small ones."

LOBSTER AND CONCH

Life got better many years later, when people started buying lobster and conch. Lobsters became especially popular in the late 1950s. They had plenty to sell.

"We caught them alive. We had a boat with a well, to bring them here to sell. We put them in a wooden box, put them in the sea. A boat from Belize (City) came to buy once a week. Still it was very cheap, but we had something to do. We started to get more money at that time."

Regina lived on the money her sons gave her. All the brothers gave her money for food and clothes.

They lived on the beach, on land owned by Sam, Secundino's eldest brother. Then Hurricane Hattie destroyed their home in 1961, so they built a house in town. Secundino's youngest brother built the house and lived with Mrs. Gonzales while Secundino lived opposite. Sam later turned his place into a hotel.

"After school I used to go fishing. We used to go to Corozal and Chetumal to sell the fish. It was very difficult; sometimes there were ten boats there. After that we went to fish for lobster. I was only 14 years old when I started fishing." They caught lobsters with a small net and a glass in about seven or eight feet of shallow water. Or they used a big steel rod to catch fish.

CHANGES

As the situation in San Pedro improved, people began building better houses. There were 15 lovely houses on the beach, but the hurricane destroyed them in 1961. Everything had to be rebuilt. Some of the Americans who came helped to rebuild the houses. They also built hotels which attracted tourists to the island.

Secundino and his brothers took tourists fishing for US $20 per day. Snorkeling also began to become popular, but with only a mask and no snorkel. The reef was much more pristine then, and lobsters were abundant. There were also many lobsters in the lagoon and on the back of the island, where there are now only a few due to overfishing.

Both fishing and tourism brought better incomes to the people of San Pedro, Secundino explains. "I started taking out tourists. I had a small boat, fishing for snapper, sport fishing. There were a lot of fish at that time. Then we had diving too. Snorkeling started in 1965 or the '70s, something like that, diving with the tanks around the 1980s.

"When the tourists came, people from all over came to

work, and the island started growing. Now there are a lot of people here from Guatemala, Honduras, El Salvador, Mexico, and all over Belize, and that is how it has grown."

A GRANDSON IS A DOCTOR

"I only recently started the pharmacy. I sent my son to study to be a doctor. I have six children, three boys and three girls. The pharmacy I opened only three years ago, because my son is the doctor. He would send a lot of patients; many people would come to buy their medicine. And there was no pharmacist."

His daughter manages the pharmacy. They order medicine from Corozal, Orange Walk, Cayo, Belize City, even the US and Europe.

"When my son was very small, I put it in his mind that he would be a doctor. He grew with that in his mind, he would be a doctor. He wanted to be, so he studied very hard. He studied in Mexico and Guatemala. First he went for two years to Mexico and then for six years to Guatemala, and then he went for another three or four years to Mexico to specialize in gynecology. And then he came back to work here. He had met a lovely lady, a medical doctor. She had spent time abroad."

ONE OF THE RICHEST ISLANDS

"The future of San Pedro will be good. If nothing happens and they don't have a big storm, in five years this will be one of the richest islands: beautiful houses, hotels, condos, a lot of people working here.

"San Pedro has a lot of investments and beautiful houses and condos in the north; the south is almost full. It is good to invest in the north; the land is more solid, the north is much wider, the land is higher and has more fresh water than the southern part. It is nice in the north, and they have lots of land to sell." (Contact helga.peham@chello.at.)

PART II, CHAPTER 15:
"A PARADISE WITH A PAST"

Mr. Angel Nuñez, the Director of San Pedro High School, was born in San Pedro in 1950 and raised there.

ANGEL'S BACKGROUND

"I left the island for my formal education, and started teaching at age 21, in 1971. I am a cofounder of San Pedro High School. I've seen it flourish and blossom from a tiny little school to what it is now, basically from my concern for young people. I've seen two full generations of graduations. My dream is to leave a well-founded school as my legacy to education.

"We offer all four high school grades, 9 through 12, and we also run a night program for junior college. Our students are roughly 13 to 18 years old. After junior college they can go to Belize University for two years to earn a Bachelor's degree.

"Our school was founded in 1971. We had 25 students, and 10 graduated. Now there are 380. We started our school at the town hall, which was also the community center. We borrowed it from the village for 12–15 years before we accomplished a dream and built our own school. Embassies in the US and Canada, NGOs, the Lions Club, and UNICEF tapped into their resources to fund our school, and here we are.

"One of the reasons for the high school was that we had an

average of two students per year back then going on to high school at age 13 while the others went on to become fishermen; that is what we wanted to change. To give a wider sector of young people the opportunity, change that trend. Now almost 100% look forward from elementary school to enrolling in high school. In 2005 they had 150 students, proving that San Pedro High School is doing its share for the young people.

"We killed the fishing industry. We took a lot of fishermen away from the sea, but when tourism started growing, they needed receptionists, front desk clerks, accountants, and hotel managers. Education closed one door, fishing, but opened another door, tourism. Junior College offers business courses and tourism related courses."

SAN PEDRO'S PAST

"When I was a child, San Pedro had a population of 500. They lived in small thatched houses with fish all over the ground floor, chickens running in the backyard, fish all over the fences. Canned fish, salted, preserved for weeks, dried, fish all over rooftops. People were barefooted. They earned their living well, a humble life not of poverty. All men were fishermen, all women were housewives.

"People married young. Women were aged 16 on average, men 18 or 20. Large families, no contraception or family planning. No construction, no television, no telephone, electricity only from 6–9 pm, no airplanes, sailboats only, no water taxis, no mail, no ice but we never missed it until we got an ice factory.

"Very few social activities were going on. There were maybe three, four parties during the year. Entertainment meant going to the bar. Alcoholism was a problem. In the '60s and '70s there were no drugs, no marijuana, no cocaine, no crack – that only started in the '80s. Population growth, other cultures came, new ways of living with new people from other areas. An attempt was made to open a brothel in the 1980s, but the women and the owners saw that it didn't fit in so they closed it down.

"Mass was once a month; it was a Catholic service. There was an Adventist church for the 1% of the population who were Adventists. No other churches were here until the 1990s.

"My parents are both from here. My father is alive. He was a fisherman, then a businessman. We had chicken and pigs. Few vegetables were grown or imported. Some fruits, like papaya or watermelon, were brought from the mainland. We ate manatees. Our source of meat was manatee, salted, dried in the sun and sold at 10 cents a pound. Here in the lagoon we still have a lot of manatees at this time. Now they are protected. It is sweet meat, tender, very delicious.

"We ate turtle meat, very popular, a lot of fish. Fish was our diet, no doubt about that, almost seven times a week. People raised chickens for special occasions: Christmas, a wedding, a birthday party. Eggs were not eaten; they were also reserved for special occasions, or special dishes or birthday cakes. Not refrigerated either. Fruits were special. When a boat arrived from Sartenea or Corozal, this was a very special occasion, everybody rushed with their bags, and they bought mostly fruits. I never tasted a tomato until I was 12 years old. I didn't like it at first. Not a common thing on the island."

EDUCATION IN SAN PEDRO

"I spent eight years in primary school here, then went to Belize City for high school, junior college and university. I specialized in science and Spanish. I always wanted to teach Spanish. In junior college we had psychology, business, testing and grading – everything required to go into teaching.

"In the earlier years here, there was only high school. The job trend now is to demand higher education. More specialization is needed.

"My first job was in the high school I founded, along with Mr. Frank Nuñez, who is no longer involved.

"The first years were challenging. We had to prove that we could be as good as the city schools. It required a lot of persistence, patience, courage. Some years, only four or five people enrolled.

"After our first graduation in 1976, students proved themselves by becoming gainfully employed, and things started changing. Then we had a growing school and growing needs. Fundraising and planning for this building started in 1981, and we moved over here in 1987.

"Some of our early teachers were local, junior college graduates who came back seeking jobs. Others came from other parts of Belize and applied, and a group came from the US Peace Corps. They usually taught mathematics and science, since we had a scarcity in Belize in these areas. Now most of our staff is local. We have 23 teachers, 18 students per teacher.

"All private schools in Belize receive government grants. These cover 70% of our administrative expenses. The other 30% are paid by special fees from students, very minimal. They don't pay tuition, but pay fees to cover 30% of the cost.

"In our computer lab there are 25 computers. We started with a small lab, maybe 12 or 15 years ago. Our computers are adequate, but never enough. We should have more space and more teachers. We can only offer the computer lab to the two higher classes in high school, the juniors and the seniors, and to the junior college. (Info: helga.peham@chello.at.)

"Plans for the future of the school include making sure there is continuity, that it will remain in good hands. I hope that can happen with a staff that is very caring – perhaps locals will help – and a principal who will be respected. Possibly graduates can take over; there are several good candidates at the University of Belize.

"Students must become productive. The success of a school depends on its students. We must produce what society wants. The government of Belize has been supportive of education. I think that people have accepted that only through education can they take care of their future, and so they go for it."

ANGEL'S FAMILY

"I have one sister and one brother, who passed away at 55. My father's family was small. My grandmother, on the other hand, has nine brothers, and hers is the largest Nuñez family in San Pedro. My other brothers work in the factory, as tour guides, in other parts of the tourism industry – all over the place. We have second-generation high school graduates in our family, working in tourism. I supported my son establishing Ambergris Today when he started it in 2000 by doing some writing for the weekly paper. He is going to prove himself.

"I have been writing for the San Pedro Sun since 1993. It was the first paper on the island, run by Bruce and Victoria Collins back then. I wrote a column about San Pedro's history 25 years ago; now this column is run by my son.

"I am San Pedro's master of ceremonies and never charge for any event. I have done over 700 in my life: over 200 wedding toasts, political events, civic events. I have written close to 30 eulogies and many other programs. I don't charge for any of those; these are only voluntary, honorary tasks.

"Members of our family have been getting more involved in community affairs. The Queen of England gives awards which honor five or six Belizeans every year. In 1998 I was chosen to receive this award. It felt good to be the first and only San Pedrano. I got this award because of my involvement in education and community affairs."

FUTURE PLANS

After he retires, Mr. Nuñez will write a book about Belize as it was 25 years ago. There are some crazy stories to be told.

"One out of every ten babies ate sand until their parents stopped them.

"Turtles lay eggs; a single turtle lays 150 eggs. They are all over the island, though it is difficult to locate them. Turtle eggs remain in a liquid state even after boiling.

"There is deer hunting on the island, wild boars, wild pigs, in the north. Wild turkeys are on the island, in the Basil Jones area.

"People used to interpret clouds. They saw kings, queens, dragons, angels. Animal lovers saw animals and nature lovers saw mountains. It took quite some time to reach Belize City. Transportation from San Pedro to Belize City was 10 hours by sailboat. So it was a nice pastime playing with the clouds."

ANGEL'S THOUGHTS

"I have been through two major hurricanes, Hattie in 1961 and Keith in 2000. I believe that this island is blessed, that it is safe no matter what others say. I have seen it go through two Category Four hurricanes, and they have not damaged us that badly because of the Barrier Reef. I feel safe here. Nobody has died here from a hurricane. When hurricanes arrive from the east, people are evacuated from the island.

"Big changes have occurred in San Pedro in my lifetime. It

was once a small fishing village with no electricity, but now it is a bustling town."

Mr. Angel Nuñez is proud to be grandfather of a little boy.

"I try hard to ensure my grandchild will be bilingual; he learns rapidly. Everybody should be bilingual. Spanish is important, a necessary tool. English is the first language; since all documents are in English, everyone should know it. Creole people should just learn Creole on their own, it shouldn't be taught, but the children should master Spanish and English first, and Creole should be a side hobby, a dialect we learn in the street.

"I love all of Belize. I still think San Pedro is the best little spot. I love Cayo. I love mountains; they fascinate me. I love San Pedro for the people. I have a lot of friends in Stan Creek. There is a growing group of expatriates in Cayo."

ANGEL'S ADVICE

"I recommend for anybody settling in Belize to learn the history and the culture of our community and try to adapt to it, not to change it. I have heard a foreigner say, 'The schools in Belize are not good, no world or US history.' I don't agree with people coming and trying to change everything. Learn about our historical roots and culture. You want to enjoy our culture; if you loved your own culture you would have stayed there. Some people have the feeling, adapt and enjoy what we have. When foreigners come to Belize and live in our society, they benefit from us and us from them, economically, socially, culturally.

"I think the growth of San Pedro was necessary and welcomed. We couldn't stop it; it had to grow from this humble village to a bustling town. This led to a rise in the standard of living; income was increased because of the population boom. We have 10 or 12,000 residents now, and maybe 2,000 more tourists. Stores, hospitals, rental – all people have found income, it was necessary and welcome.

"Population growth brought some unwanted unwelcome changes that we had to learn to accept, such as lost freedom. The children used to play freely on the streets, you could leave your house open, you were sleeping with windows open. There's also loss of privacy, and children are not 100% safe from drugs and criminal activities.

"There is a fear for loved ones. To welcome growth one has to expect these evils, protect against or try to change these evils. We can as a community reduce crime, theft, drugs, and AIDS with a unified education, not only in schools but also families.

"Smile at the progress, lament some of the freedoms we have lost."

PART II, CHAPTER 16: FOUR YOUNG PEOPLE AND A NEWSPAPER

Dorian Nuñez and Perlitta Zabata are Angel Nuñez's children. Dorian initiated and owns the weekly newspaper 'Ambergris Today.'

Learn more about the newspaper by contacting helga.peham@chello.at.)

Dorian Nuñez

"I grew up here in San Pedro. It was small and quaint back then, quieter, with less traffic. I went to San Pedro High School, then went to St. John's College in Belize City for sixth form. I earned my Bachelor's degree in Journalism from Benedictine College in Atchison, Kansas, in the US.

"Now I own a newspaper in San Pedro. I founded Ambergris Today on February 4, 1999, when I returned from college. It was a challenge. I actually worked with the other newspaper first, San Pedro Sun, but after a few months I thought I could do it myself and I had a vision of doing something different, something that was my own.

"I got the equipment that I needed, the basic office stuff. The hardest work was to go out and introduce this new newspaper, its goals and how it would help the island. My goal is to have a newspaper with a positive attitude. We all keep that in mind when we write articles.

"What we want to achieve with the newspaper, our main purpose is to inform and educate our community and let them know what is going around in town, assisting our community in every effort that we can.

"Our staff is four people. Every week we go out to get information from our sources. We do the whole layout here in the office. The writing, the advertising, all the layout is done here and we print in Belize City. Norman's Printing House accommodates us very nicely. It comes on the first plane in the morning. We deliver it ourselves, keeping in contact with the people. It's fun to deliver to all the businesses, nice to get out there.

"Edouardo is our main writer, and also Nelly a little. We share the contributions. We talk to each other a lot.

"We archive our old hard copies. Since 2000 we've also archived back issues on the Internet, on our home page.

"We saw growth in our newspaper this past year. The business community is growing. There's more advertising. Our main goal is to meet the demands of this growth this year, especially with our website. More people are visiting our website. We are adding more informal information, articles, columns, and features. Mainly we focus on San Pedro, but we also get information from the mainland which our people here are interested in.

"My work is fulfilling, you help and inform your community; your product is out there to see. If you like your job, you are fine. Every week people's feedback will tell you if you did a good job. The other thing that I like is that it is changing; there is not always the same thing happening every week.

"I am single, living in the San Pablo area. It is nice and quiet, a lot more tranquil than in town. A year ago I moved out from my parents' house to my own house.

"The main business opportunity is definitely tourism. If the people come and embrace the island, people will embrace and support the businesses. As needs arrive in everyday life, we find ways to provide for them.

"Real estate is a big business with a big future. The fastest growing business on the island is land speculation, raw land and development. Raw land, found just as nature made it, can be developed if you put a house, business, etc. on it.

"Many people retire here, and we advise them to study, do research in the community, on the environment, what kind of communities they are interested in, such as Cayo, Belmopan and Corozal. Corozal has a free zone, which is the main catalysis for growth; the free zone will grow. Business in the free zone is also a good place. In Cayo expatriates own hotels, restaurants and resorts; contact them.

"One of the main things seen here when someone is happy retiring here or owning a business here is when they embrace the community, make friends, care for the island, take care of the reef, the land all around. They stay longer and are happier here, and we accept them into the community even more. It is a great way to get to know people, to be accepted."

Edouardo Brown

Edouardo is a reporter for Ambergris Today. He was born and raised in San Pedro. His childhood was filled with adventure and freedom. The island was a small fishing village where most people knew little about the rest of the world. Laws were few and Edouardo had many ways to express himself.

"It was my greatest experience in my life; it was really a paradise on earth. I grew up right here in town; it was a village then.

"There was only one policeman in the village. He didn't work a lot because there was little need for him. There was no crime, no drugs, just too much drinking at feasts sometimes. He registered births and did other documentation, and he probably acted as a Justice of the Peace.

"I was 17 when I finished high school and went straight to the US, to the University of Arkansas in Little Rock. I majored in English and minored in legal studies; it is called a Liberal Arts degree, a Bachelor's degree. A friend assisted me and my father also paid for my education. My father is a tour guide, one of the first tour guides on the island. He is still working."

In 1989 Edouardo earned his degree and returned to Belize. He started working as a tour guide and for many years did a lot of scuba diving.

"I started working at the newspaper when Dorian opened it in 1999. The island is small; basically everybody knows everybody, comes into contact with everybody. It was very compatible with my person. I like reading and writing a lot."

Edouardo enjoyed every aspect of the newspaper: meeting all kinds of people, covering all types of events and learning photography.

"My job is the reporting. If we are called to do an article, a piece of reporting, I am usually the one who goes there and gets the information and writes up the article." By staying in touch with the community they learn what is scheduled for the following week, and people also call the paper to report current events. "We always maintain good contacts and give excellent service to the people we meet. Being very frank and very patient, while knowing how to have good public relations with people, helps your business. It helps both sides; in every business it helps."

Edouardo's view of the tourists is favorable. "Personally I see the tourists and the expatriates enriching the lives around. In every way, you take the positive from the people or cultures because if there is any negative this will undo itself without any help from us."

Edouardo's family has always lived on the island. He, his wife and their daughter live in town, and Edouardo walks to the office.

"The future of the newspaper I see as very positive, a learning experience. Whatever happens, we learn to make it more positive. As for the future of the island, I see more development with more hotels, more businesses and more people, unless something happens. If a hurricane comes here, that would be a setback. But I still see the island growing more and more. Thus far we have done a very good job of keeping its beauty."

Ambergris Bay will be a planned town on the west side of the lagoon's north island. It will have wide streets and a good water system. "It could be much better than San Pedro town. It is very likely that it will happen, because in San Pedro town there is not much space for expansion. When this will be built is anybody's guess; it could happen next year or in the next ten years. This depends on the people living in town, those that have land titles – many have titles, some have 100 or more – and the government of Belize. Mostly people from Belize, but also other people, will get land titles in Ambergris Bay."

There are several newspapers in Belize, many radio stations and a few TV stations. The press in Belize is very free. "All newspapers in Belize have their good sides and their negative aspects. I believe most of them are privately owned. For example, Belize Times belongs to the People's United Party (PUP), The Guardian is owned by the United Democratic Party (UDP). All other newspapers do a pretty good job of staying out of politics.

"I see very positive things for Belize, especially as women are running for political offices and they are getting elected. And Belizeans on the whole are being less tolerant and more outspoken against negative things and corruption. The main reason for this is because Belizeans are now more educated than ever in the history of this young country. St. John's College is excellent, but now we also have the University of Belize (UB), which follows in the footsteps of St. John's College. What they study at UB prepares people to do legal studies, to study medicine, to be doctors.

"When people come here, they have to find out everything

that they need to know about Belize. So don't rush in. I would suggest before you come to live in Belize, if you can, live in Belize for three or four months, and if you can six months, and see if you like it. The people are very nice but the culture is very different. And in that way you can really know if you would like to live in Belize. If you like a country with lots of adventures and that is a little rough around the edges then you will probably like Belize.

"I have been blessed to have the opportunity to live my life on this island and in this country."

Perlitta Zabata

"Being a kid here in San Pedro was great. Many neighborhoods were good, not much traffic outside, in the street in front of my house. Primary school was good, teachers were all locals, and it was nice. There was the Roman Catholic school and the high school, San Pedro High. After high school I started working at Tropic Air in Operations, in the Reservations Department, basically telling people the places to visit, making reservations, dealing with people in the main office. It was fun; every day you speak to your friends like the hotel clerks, and meet different people."

Perlitta graduated in 1996.

"I started working at the newspaper part time when it started in 1999, then in 2001 I started to work full time and I quit Tropic. The newspaper was expanding its size and circulation.

"My job is graphic designer. Self taught, good help from a

cousin of mine who is a graphic designer. I like anything that has to do with art. It is like playing, because it is fun. I've worked in advertising, but now Melody is in charge of that.

"What is needed depends on the client. For example, the owner of the hardware store wants pictures, so the reporter goes out and takes the pictures with a digital camera. We have a lot of job opportunities here. We have a lot of people coming from outside. The island is expanding. There are more investors than employees. I see business opportunities are improving because more people are getting interested in Belize like they were for Cancun. I see the island's future as a mini Cancun.

"I think it is going to be better, a lot more movement and more jobs. More investments will come and there will be more people interested in coming here, and tourists coming for diving."

Perlitta met her husband at school. "We got married on March 16, 2002. We had a wedding in a church and the celebration in the high school. They use the high school a lot for celebrations. They have a big yard and a tennis court.

"I have one child. He is 3 and in preschool learning the alphabet, shapes, numbers, colors, getting ready for the 4-year-old preschool, socializing. It is a private preschool, ABC Preschool. The owner is Will Alamira, a Belizean. In my boy's group there are almost 30 students. School starts here at 5, like in England. After Easter they evaluate the children in preschool. I also teach my son at home. I give him a head start.

"I would want people moving here feeling at home and taking care of San Pedro. I think this would be the strongest thing

– love the island as we do. San Pedro is different to the US. Just be open-minded because we have a different culture. We have Mayan Garifuna, Creole, Mestizos, Mennonites, Chinese – we have a lot of people here. East Indian, Lebanese, many others.

"The next government is going to change. I wouldn't be surprised if it changes. Change is always good. When a party stays too long, they start doing stupid things. Power gets to you after a while.

"Things have changed a lot from my grandmother's time. Most women stayed at home then. Now women work. A lot of women work – more than 50%. There is a lot of work on the island. Often females are preferred for jobs. Maybe we have a better character, maybe we are more flexible than men. It just depends on the attitude you have at work.

"We have to adapt to the foreigners who live here. They have to adapt to our culture. Improvement and money are the good side of foreigners – investment, jobs, and knowledge."

PART II, CHAPTER 17: SURROUNDED BY CHILDREN – ISLAND ACADEMY

Wilema Gonzalez was born in Belize City on August 5, 1973 and raised on the island, where she has lived all her life. She went to the Roman Catholic primary school, then to San Pedro High School. She had a simple childhood because there were not so many cars, resorts or people, so she felt like she was in a big family.

GROWING UP IN BELIZE CITY

"We were able to just wander around in the streets, play with the neighbors, nice and simple. I enjoyed it most when it felt free; I didn't have to worry about anything or anyone harming me. We went out in groups to local dances."

Wilema's father is a fisherman and her mother is a housewife. She has two older brothers, Eddie and Will. Eddie is a dive instructor in Amigos del Mar. Will is a preschool teacher.

After high school Wilema worked at Victoria House for two or three years. Then she went into banking, starting as a teller. "I worked my way up to supervisor of customer service and operation supervisor, five years at Belize Bank, three years at Atlantic.

"Always it felt like you had to be extremely careful, because at the end of day everything had to balance. It was very stressful. Much responsibility, confidentiality and accuracy was needed.

"When I worked there, there were only two banks on the island. I felt they were pretty much the same. Belize Bank is owned by Michael Ashcroft; Atlantic Bank is owned by some Honduran company. I was in charge of local operations.

"At Atlantic Bank I met my husband. I had a high-risk pregnancy. I was obliged to resign because we weren't allowed to work together after we got married. I had a healthy baby. I stayed nine months at home, and then I went to work for Dixie Bowen in the Island Academy. Later, I had my second baby."

THE ISLAND ACADEMY

Wilema started working for Dixie Bowen (Miss Dixie) at the Island Academy in 1999, half a day in administration and half a day as a Spanish teacher. After a year she quit teaching Spanish and stayed in administration.

"At the Island Academy every day is different. You are busy and at the end of the day you know you have accomplished a lot. There is always something happening here. Most important, a good academic education is great. If you don't have good manners, or etiquette, you will not go far. Therefore discipline is important and so are good manners. The children learn how to present themselves well. The Academy wants to make a complete person out of a student.

"I wasn't here the first years. From what I have learned,

Dixie always had this vision in mind. Her two kids were in primary school, and this gave her ideas. There were three buildings and 70 kids in the beginning. It needs an average enrollment of 55 kids in order to maintain itself. This is a tuition-based school. Tuition covers salaries and utilities.

"Mr. Bowen takes care of the maintenance of the school. Dixie is principal and owner, and she does not take a penny out of the school. Dixie felt there was the need for such a school. The primary schools were too crowded and still used corporal punishment. We do not touch our kids. There are other methods of discipline.

"There have been six classrooms since the beginning. We are happy to keep it at that, with up to 15 kids per class.

"The children are from business families, the ones that own hotels, dive shops, tourism agencies. We have Lebanese, Americans, Canadians, South Africans, Belizeans, had Chinese in the past, and British. The teachers are American, Canadian and British. There is one Belizean teacher on staff. We advertise on the Internet and get resumes. We also recruit locally.

"Sports are done on the premises. There are no computer labs for the students, but we are trying hard to get one."

"I am planning to stay in the school as long as it is here. I would like my kids to come here. I feel very useful here. This school gives me the opportunity to reach out. I will leave only if my husband has to be transferred."

PROGRESS

"I know that progress is good. It has helped the country a lot. But I see Belize developing so fast that it's depressing. There is the beauty of the island, natural beauty. You need to control certain things. The traffic is getting out of hand. Everything is growing, developing so fast, in the end we will be losing money if we do not control the growth; we may lose what we have. People see money, more money. If we do not take care, all this building will destroy the beaches. We need to set controls.

"Two things about growing up here. A lot of what happened in town revolved around the church: Easter functions, Christmas, rosaries, Posadas, the bible story of Christ's birth, prayer groups. Christmas felt like Christmas. And everybody would go to everybody's home; it was very family-oriented.

"I cannot even explain how beautiful it was when I grew up here. There were more coconuts, more beaches; it was much better when it was just a little village. I would have loved my kids to grow up as I grew up. Now I hold their hands for ten minutes to cross the street. There were only two land rovers here when I grew up.

"You want more and more, but there is a point when we need to stop. We have to realize that, to set limits and controls. What I have experienced personally is that, because of all the modern things we've adopted from other places, we end up complicating life with so many things. We want more; we don't realize that we complicate our life. Before, you learned to live with what you had. We have forgotten that we once used to live here on the island and we were happy.

"I was happier with a simple life. I never starved, I had what I needed. We were on the beach; it was one big family that took care of each other."

PART II, CHAPTER 18:
A DILIGENT HIGH SCHOOL STUDENT

Juanello Grimaldo is a young San Pedrano. "I used to sell food in the street. All my family used to sell in the street. We woke up very early to sell in the street, all around town we used to sell. When school finished, we used to sell again.

"We used to sell sweets: salted bread, sweet bread, sweet flan, maha blanco [white yoghurt blended with rice]. My mother made all this. My mother used to work hard when we went to school.

"I sold this with my two little brothers and my two little sisters. I started to sell at eight years old. I went alone and my next brother went alone. I used to sell a lot. I loved it. I used to make $55 a day.

"I stopped when I was 13 years old, because I found a job at Lilly's then. I was a waiter, same like now.

"In October 2004 I stopped at Lilly's and went to work at Amigos del Mar. My job was to fill [scuba diving] tanks. I used to meet a lot of people from the US, from Colorado. I have a friend in the US, his name is Bob, a big tall guy. He likes to play soccer like me. Good to work at Amigos del Mar, they treated me well. I was one year in Amigos then I quit again. I felt tired of the work. I came back to Lilly's December 26, 2005.

"Now I work and go to school. I go from 6–10 to high school, third form.

"I am going to St. Peter's College. My favorite subjects are Mathematics, English, Accounting and Spanish. I go to play soccer, football. The school's football team is St. Peter's College football team. The school is at the Boca del Rio. The principal is Mr. Cobb, from Corozal. Vice principal is Mr. Hernandez, Corozal. The owner of the school is Mr. Frank Nuñez; he has a son, a teacher too, Alex Nuñez. We have a new building; we are currently building the second floor. The students from the third form are building during the day. We learn to build houses.

"I live together with my parents. All my sisters and brothers are at home, we are five all together. I have in total 10 nieces and nephews, that is a lot. I am the youngest. I help my family too. I give them some money, and stuff like chicken, I take it home for lunch.

"We played games at home with the family, with my sisters, brothers, my mother: dominos, bolotti [card game played with 10 people similar to Uno].

"My father used to go fishing; he was a fisherman. He was diving for lobsters and conch. I used to go with him fishing and diving. We took the things to the cooperative. We are member in the cooperative. My father is still fishing and diving for conch and lobster. We still have lobster although not so much as before.

"We went fishing in the lagoon. My father went outside to Turneff Reef. He used to go there for 15 days with his sisters and brothers and my mum. My mum used to be a diver too.

"All my brothers went to sixth form in school, there is only me left to finish school. My sisters finished fourth and sixth forms. My two sisters are going to university in the States in Colorado. They got scholarships through the school. I have a scholarship too. I play soccer, football and got a scholarship. All the players got a scholarship – we are 15. The people who give us scholarships are from Colorado. Their names are Gregg and Cory. They are nice people. When they came here for vacation I loved them. For two years now there are scholarships for all of us. They brought some coaches and some cheerleaders too, young ladies. My two brothers are working. One is police officer in San Pedro, one is a BDF [Belize Defense Force].

"After finishing school I want to work in a bank. I want to go to university in the States. If I still have my scholarship I will go to the States. Or maybe I want to be a pilot on a big plane. I want to fly planes from Belize to the States to Europe too. Maybe have my own resort, one day. There are a lot of things that I can do. I will look at all the possibilities.

"How did San Pedro change? There is a bigger park now. There is a medical university here. Also there is a bridge to the north. They have a new basketball court. They paved the streets. When Hurricane Keith came in 2000, the coconut trees and the coconuts fell. Rooftops of the houses and the ceilings were flying. Our house dropped, nobody was hurt. The water went to five feet in the house. We were in a shelter in my uncle's house. It is a cement house and it was safe there. We fixed our house back again all over – my two brothers, my father and me.

"It took three months to clean the entire beach and the

streets. Palm trees were all down, a lot of things, houses. Many people were evacuated. When all the damage was fixed the tourists came back. Keith was in October 2000.

"We helped the government to clean the beach. They needed some workers.

"What is the future of San Pedro? This island here used to be small, the rest was bush. I remember that my grandmother told me that. They will build more. They are starting to build a new school on the other side of the river. I will stay here in San Pedro or study in US, maybe fly around the world if I become a captain.

"My friends are good friends; they are not gangs hanging around, they are from school. All my friends are playing soccer/football. We are going to Virginia next month to play soccer. It will be the second time for us to leave and play.

"I have a girlfriend right now. She has finished school. She works in the Atlantic Bank. I met her when I went to the bank. She is a teller. I like going out with her. The Club in Jaguars and Big Daddy's is where young people go when they go out.

"The tourists are kind of friendly. They bring the money to San Pedro from the US. There are tourists from a lot of places. Some come from Europe, even from Iraq, one came from China."

Looking over the Fence

Belizean Boy

The Hand-Driven Ferry, before its Shut-Down

House in San Pedro

Houses on the Lagoon

Pelicans Looking for Fish

PART III: REAL ESTATE

PART III, CHAPTER 19: TIMESHARE – SHARED TIME

"Previously, I had never really heard about Belize. When I was visiting my ex-boyfriend in Greece, another friend wanted to work for Captain Morgan's timeshare in Belize and needed someone to join him. He told me about this place, and we went together. So I came and I never left. Timeshare is a really great thing."

TIMESHARE IN SPAIN AND THAILAND

Priscilla van Ameyde has been working in timeshare for 12 years. She started in Spain with outdoor personal contacts.

"You just phone a timeshare company and ask, 'Do you need any workers?' They pay the plane ticket and a month's free living. I started working in Spain. I was working on the mainland, the Canary Islands, Ibiza, everywhere for four to five years. Then I worked in Phuket in Thailand where I did the same thing.

"Basically I approach people. I talk to couples and ask them if they want to go to a timeshare presentation, and then at the presentation the salespeople try to sell. I get a commission on every couple that goes in. If they close a deal you get an extra remuneration. You need to talk a lot, you need to convince people to do what they do not want to do. I make a lot of money.

"You can travel all over the world with this job. I phoned Thailand and stayed there for 8 months. It simply was great. They have the best food in the world. They give you an apartment. This one here in San Pedro I got myself. You say 'listen' to the people, you just tell them a story and they get interested and go to the presentation. Many young people travel around the world this way.

"You have to know how to sell the units. In Spain, Thailand, and India, there are interesting people, different cultures you get acquainted with, and you have the time to do something else. All are 4- and 5-star apartments in timeshare. But I've also seen so many fail in this timeshare business. You have to get the people to like you, and do what you want them to do. It is a lot of personal liking.

"It is pretty interesting, especially when you are young and want to see the world. I got so many job offers because of it. Sometimes I earned €2000 a week and spent it by Friday. The best was €9000 a week in Spain. I made so much money and spent it on clothes. European clothes are better than American."

EARNING GOOD MONEY – TIMESHARE BUSINESS IN BELIZE

Priscilla worked for Captain Morgan's timeshare business for one month. After that she didn't work for half a year, just happily enjoying a holiday. During this time she went to Mexico. Then she started working for Banyan Bay, another timeshare company in San Pedro. She found an apartment within one month of coming to San Pedro, but there were mosquitoes everywhere.

"I have to leave quickly, I thought. Either I take an apartment on the beach, or I leave. Then I moved to my gorgeous little apartment. After a month, you get adjusted. Now I know a lot of people on the island, and everybody smiles. This is a pretty happy island. I pass my time by hanging out. I go to bars, drive my golf cart, meet people. You have to be an open personality."

OWNING A VACATION RENTAL BUSINESS

Priscilla and Michelle soon became friends. Later, they opened their own vacation rental business.

"We came up with this idea. Michelle did the writing and I did the photos. It is starting slow. American and Canadian holidaymakers are our expected customers. More straight flights to Belize are needed. Europeans travel a lot, but there are no direct flights from Europe to Belize."

Priscilla thinks Belizean food is probably the worst diet in the world because she does not like burritos and tacos with various fillings.

"I like the people, because they are outgoing, friendly, funny. I like the little children, they are really cute. Everybody here drinks. So it is nice to hang out in bars. Like in Fido's — you know everybody. You know one another, you do business here."

TRAVELLING

Belize doesn't have much export and import business, so many people go to Chetumal in Mexico to buy things.

"It's great being here if you want to see a lot of South America. It's close by. If you want to go to El Salvador or Costa Rica, you are there quickly."

Priscilla spent three days in El Salvador, which she enjoyed, even though many people told her it was dangerous. The beaches there are covered with black sand, and the waves are the largest she has ever seen. Also, she says, it is very cheap there. She stayed in a little hotel that wasn't in the commercial zone. One of her advantages is that she speaks Spanish.

"The landlady sent me with her son for food. In the evening I went with a taxi driver to all the bars. He took me everywhere. That was fun. It was dry on the beach at the end of summer, just before September, OK for a couple of days."

Priscilla also visited Tikal, in Guatemala, and her impression was of small, cute shops. And an unexpected problem. She wasn't able to use her debit card, so she ran out of money and had to leave earlier than she'd planned.

"I had to get money sent. Next time I will bring enough cash. Tikal is lovely, with huge trees. I have never seen such special trees in my life. Flores is a little Guatemalan village where you can buy clothes, skirts and bags in funky colors at cheap prices."

Priscilla also enjoys going to Mexico.

"My ex-boyfriend and I met in Playa del Carmen. I went

there a lot. We always went to Tulum, which is very special, a nice place to be. Little beach cabanas are only US $10 or $20 there. San Pedro is extremely expensive compared to this. I used to pay US $900 rent for my apartment in San Pedro. Now I pay US$600 for it."

BUSINESS IN BELIZE

"Sometimes people are so ignorant when they come down here. They should just be open for anything. I hang out with the homeless, the Rastas, and the owners of the resorts.

"Everything here is a hole in the market. Everything is a new thing here. Magazines, newspapers, phone company. They have nothing here. A friend is starting a purified water system business for US $3,000 with no competition. If you have a mind for small business you can do a lot.

"Real estate is another good opportunity. My neighbors are building on the other side of the lagoon, opposite the boatyard, on the small islands. It looks European, modern. It will look like Cancun there.

"If you have the money, develop. The North will be big. There is an airport, an airstrip there. A good place to invest is in the northern part of Ambergris Caye. Now they also have the bridge to the northern part of the island."

A TRUE STORY

Two young women were smoking hash at a private party at home.

"One went out to get something when a police officer stopped her, checked her and found that she had been smoking hash, which is an illegal drug. The police entered the house and put the two women and a man into jail.

"They got out of jail with the help of a lawyer friend. Actually, he told the three young people that the police are not allowed to enter a house without a written warrant from a judge. The police in Belize are not well paid and are corrupt. You have to know your rights.

"Weeks later the two young women danced in a disco with the policeman who had put them in jail and they became friends. Also that is Belize.

"A big problem in Belize is the lack of education. Some people are therefore so primitive because of lack of education."

LIFE AND PEOPLE IN SAN PEDRO

Priscilla loves a lot about San Pedro.

"I like the atmosphere here. Nobody gives me any problems, I don't give them problems. People are really helpful. I would definitely like to retire here. It is a great place for retirement. You can lead an uncomplicated life here. I left Holland because there are too many rules. This was my major problem.

But there are also some people that come to this island and they get deported.

"There was a man. We took him to church twice in my golf cart. I remember the last time we talked, for half an hour at the supermarket. Then one day the FBI took him. He was wanted for bank robbery, so the FBI deported him.

"There are undercover cops on the island. People around here who don't do anything are undercover agents. Another guy working in a resort up north is from the FBI.

"It is relatively safe here in San Pedro at night."

PART III, CHAPTER 20:
WITH A SMILE ON HER LIPS

Michelle Kachur was born in Gary, Indiana. There are steel mills, steel companies, and other major industries right on Lake Michigan, "not at all pretty like this." She attended Indiana University on a free writing scholarship and earned a Bachelor's degree in English Literature and History. Her mother was an English teacher. Michelle likes to read and write.

A HOLIDAY

"I came because I was working very hard in Chicago. It was January, and my boss gave me an extra week's holiday. So I found a book on my shelf, from a boyfriend in high school, 'Undiscovered Islands of the Caribbean.' I opened it right to the page where it said Ambergris Caye, and I went there. It was my first holiday alone. I had always gone with my family and my friends. This was 14 years ago. I was 25 or 26.

"When I landed, it was dark and raining. When I stepped off the airplane I felt that I would never be leaving. I felt like I was home. As soon as my foot touched the ground, I knew I would stay forever."

MOVING TO SAN PEDRO IN THE NINETIES

Michelle was fascinated with the place. "I went home, sold everything and moved here three weeks later. All people I met on the first day, I am still friends with, really nice people, from Europe, from Honduras and other places."

When Michelle came to live in Belize, the island was quite different. San Pedro was very small. Apartments were not so expensive. There were practically no cars, only golf carts. You could live on the beach. It was more like a fantasy then.

Michelle lived at Paradise Villas when she first arrived in San Pedro. After a while it became too expensive for her, so she soon found a small place on the beach.

"I had to carry big buckets of water to my hut. There was just a little place for cooking. But I had this gorgeous view. The view was the same as from Paradise Villas."

Later Michelle got a very good deal for a condo. The owners needed money, so they had to sell it for less than it was worth.

"I never worked. I had my own money and I helped a friend to open a restaurant. It wasn't easy to find a job, so I helped people do things around the island. I spent a lot of time writing stories. That is my biggest interest, writing.

"If I'd had the money ten years ago I would have bought all the real estate and we could sell it now. I think real estate prices are still going up. You can still buy land here. It seems expensive, but it is not at the top yet. How do people live here? It is so expensive."

CHILDHOOD AND EDUCATION

Michelle is Slovakian, German, Croatian and English. "In America it is hard to say you are Irish or Austrian, you are so much a mixture. My grandparents had to elope. They weren't allowed to get married because they were Croatian and Slovakian. Maybe because of religion?"

Michelle's parents have been married for 42 years and still love each other. They got married when they were 16 and 17. Her mother was pregnant. Michelle's mother is only 58 and looks 38. She exercises every day and eats only healthy food.

"I got all the bad habits of the family."

LIFE NOW

Now she has started painting. The idea of going back to work in the US does not appeal to her. When she was first in San Pedro, her boss, Neil, was her boyfriend.

"He proposed to me in Monte Carlo three years ago. So I was engaged to him, but when people asked me when I would get married, I felt like fainting. I thought that this was not a good sign and didn't marry him. We are still good friends. He is lovely and kind and a smart guy.

"You cannot hide in Belize, because it is so small. People are so forthright here. Say you came down to Belize, you cannot have an affair. People will know immediately.

"But you really can figure out who you are. The weather is challenging. It is expensive in San Pedro compared to the rest of the country. Statements are incorrect at the bank. I really want to be happy. Priscilla is a good influence. I didn't care about working, but now I want to work. I was talking to someone doing the same business we are doing but didn't want to go up to the northern part of the island. There is more than enough business around."

Priscilla and Michelle rent out luxury apartments and houses on the northern island of Ambergris Caye.

"I am working at Banyan Bay. They are training us. We are going to sell the timeshares, or at least try. I can try. I am more the girl that gives away. Priscilla is more the selling type. Money seems to come to me in life. I am more worried about people, not about money."

Michelle says that you can see fabulous things when you get up very early.

"You see the dolphins in the morning, around 4 or 5 o'clock. They don't like boat motors. As soon as the motors start they leave. If you are a sailboat, they follow you. We went to Caye Caulker on a sailboat, we tied a rope in the back and the dolphins came to swim with us. That was in my first week on the island. I loved it. You see such beautiful birds here. There is a big osprey nest up in the North."

EDITORIAL WORK; CORPORATE REAL ESTATE

After university Michelle moved to Chicago. There she worked in publishing and real estate. She first worked in publishing at Irwin and at Probus. Both companies are business oriented. Probus was built by a person from Irwin. She worked in the editorial department, but then she decided, "No more publishing for me, the sales people make all the money."

Michelle moved into corporate real estate. There she worked in marketing, doing presentations for major proposals, until she grew restless. "I could never sit at my desk. I'm not the type of person to sit at a desk all day. I quit that job and came here to San Pedro in Belize."

Michelle only worked for three years. "I was engaged, never married. I am the only one of 27 cousins not to get married. The others all have children; it is a big family."

HURRICANE KEITH

"Hurricanes are just a nightmare, the weirdest thing. Hurricane Keith came in 2000, on a Sunday. There was no advance hurricane warning. They only talked about a tropical depression. My mother's mother had just died, and she was my best friend. I was standing there in Belize when she whispered into my ear – I heard her voice – 'Go home for your birthday.' Then my friend Neil phoned from Florida on Thursday. I went to Florida on the last flight. I know how to listen to little things. I listened. This was one of those little things."

Michelle left her apartment and called a friend. Forty people stayed there for weeks; it was the only apartment that wasn't damaged.

"We lost more people than the officials admitted. They never reported the people from the back of the island. Eighty-nine body bags were flown out by Tropic Air. There must have been loads of dead. Many people were living in the back and still live there."

When Hurricane Keith struck, it first hit the back of the island.

LIVING IN SAN PEDRO

"San Pedro is a beautiful place. I think it is the nicest place for children to grow up. They just need a conch shell and a fishing line. My friend and business partner Priscilla and I pay for school for two girls. We buy their clothes and pay for their schools. It is good to help people here."

Michelle also gets help from her friends.

"Once I was really ill. My neighbor, Maria, boiled me a pot of sticks, leaves and herbs. It tasted terrible. She made me drink the tea and gave me a massage. She takes care of all her grandkids. You have to talk to the real people. The thing here is, it looks idyllic, but it is a very difficult place to live.

"It is very expensive and it is small. It takes you six times as long to go do something like getting a driver's license, just because the guy isn't there. If the office is open from 9-5, that's when the guy should be there.

"I think it is a lovely place to live, but it is not good to have a committed relationship. People will sleep with your boyfriend. They come here on a holiday and leave after a few days or a week or two.

"People drink. I want to go to the theater, but everything revolves here around drinking. There is a lot of alcohol and drug abuse. Since the army left, drug smugglers have come in. A lot more crime is here and a lot more wealth. If Belizeans have a big house and a big car they often are into drugs."

There are also people hiding in Belize. "So many people are taken by the FBI! There was an American, a really nice man. He was here for a year and I once drove him to church in my golf cart. He had robbed a bank. He got picked up. The drug dealers should be picked up. They are destroying the island.

"A woman I knew was in trouble because of drugs. This woman wanted to move back to America. She had a Belizean husband. Another woman, Daisy, took her in her sailing boat to Capricorn for lunch. Daisy sold her illegal drugs. The lady went into the toilet and didn't come out. Daisy went to check on the lady 45 minutes later. The lady came stumbling out, fell down at the end of the pier, and started convulsing. She had probably been injecting herself. Daisy put her on a little speedboat and took off and was so nervous because she had sold the lady cocaine. She went to the reef and turned. She was in a panic. The lady was already covered with flies. She took her to the Lions' clinic. Then she took her to a restaurant. The lady was dead. Daisy felt terrible. She wrapped the lady up like an Easter basket and put her in the freezer. This was the end of the restaurant. Nobody wanted to eat there. Daisy moved to the States, got married again and had four kids."

Michelle doesn't want to be alone. "I met a nice man wl lives in Florida. I met him a month ago. His wife cheated on him, so he sent her away. He lives here, runs the outside bar at Big Daddy's. He seems nice. He's Cuban. He left Cuba when he was 11. He seems pretty sincere. He's my age. He is OK and he is smart. It's hard to date someone and then not date them. Sometimes at Fido's, I see so many ex-boyfriends. There are some beautiful women here, some beautiful girls."

BUSINESS OPPORTUNITIES

According to Michelle, there are many business opportunities on the island.

"Anything in urban development would be a chance. More schoolrooms are needed as well as medical doctors. The economy goes up, the number of children rises. There is not enough of anything. Restaurants, real estate marketing, everything could be looked into. San Pedro needs so many things. My dream is that I really want to help to build a community center where kids can go after school, play soccer, baseball – a safe house. This would be good at the back of Costa Maya. People could volunteer, like at the YMCA, maybe."

Another possibility for business is to go into production. "Cashew is a fruit. They make wine out of that fruit. The nut is outside of the fruit."

Barry Bowen, the richest man in Belize, is not accepted by everybody. He is a rich and outspoken businessman. He also has done much for the island by financing a great part of the new bridge at Boca Del Rio, to the northern part of the island that was opened in 2006. His success is just one demonstration of San Pedro's many investment opportunities. A growing town with many people also needs services.

"We don't have a counselor, a clinical psychologist, or a psychiatrist in the middle level. With the drugs, the cheating, the drinking, this would be necessary. You can see how people get crazy.

"Making the front street free of cars would be good. Cars should only move in the middle and back streets.

"I just wonder what it will be later, in 10 years when I will be still here. They really should cut on the number of cars. Caye Caulker is lovely and quiet. There you will find only a few golf carts and no cars at all."

PART III, CHAPTER 21: FROM HEALTH TO REAL ESTATE SERVICES

Bob Hamilton is a real estate agent who had prior experience working in Belize. He has several university degrees. He was born on Prince Edward Island. Both sides of his family are from fishing villages on the east cost of Canada. He has two brothers and one sister.

"My brothers and I are the first generation not to fish commercially."

Bob's father joined the Air Force during World War II, where his job was to repair instruments and electronic equipment on planes. He received his education in the Air Force, and stayed there until he retired in 1968.

Neither of Bob's parents smoke or drink, and all members of his family are well educated. His elder brother has two Master's degrees and a Baccalaureate.

UNIVERSITY EDUCATION AND CAREER

"We grew up on Air Force bases at various locations across Canada. To date I have lived in eight of the provinces and I have been to all the provinces, even Nunavut."

After graduating from high school at 18, Robert earned a Baccalaureate in Business and Commerce from Dalhousie

University in Halifax, which is one of the largest universities on the east coast. He graduated at 24, then took two years off to make some money and pay off his student loans.

"Whenever I change jobs, I go back to university to get more education. I have diplomas and degrees in Municipal Management and in Health Services Management, which are three-year programs, and a three-year Computer Science diploma. They are from the University of Prince Edward Island and the University of Ottawa, Ontario. I always like to learn. While I worked as a town manager, a municipal administrator, I went to university to get a Public Administration Manager diploma.

"From there I became CEO of the Hospital Association of Prince Edward Island, a job I had for seven years. During that time I earned my diploma in Health Service Management."

LE CAFÉ DE SOLEIL

"I was in Halifax, Ridgewater, then Charlottetown on Prince Edward Island. I owned a restaurant and bar, the Café Soleil in Charlottetown, for seven years. I managed the restaurant from 1991 through 1996."

Bob had always enjoyed cooking. His mother taught him how to cook when he was eight years old. Bob made bread pies. When he was 13 years old, he was able to make anything he wanted in the kitchen. "We always had homemade bread, so I baked it. My mother was too busy."

Bob learned many of the skills of being a chef in his own

restaurant. He rented the place and built up the business himself.

"It probably cost me $100,000 Canadian to get started. The business was constantly growing. First I started out as a bakery and café and deli. That was so successful that I leased the other half of the building. There were 60 seats in this fine dining French style restaurant. A Frenchman from Dijon who had worked in Dijon and Paris was my chef. I stole him from a Halifax restaurant. He wanted to come to Prince Edward Island because it was such a nice place. I was the owner and the sous-chef. I learned a lot from him about the commercial aspects of the business. In 1998 I sold the place to facilitate my divorce from my first wife, who got 50% of the business."

Y2K

Robert went back to school at the age of 46. He took a four-month course in COBOL programming. He had done some programming in the '70s and earned a degree then. After this refresher course, he went to work for an IT company called CGI during the Y2K preparations.

After two years of this, he started programming commercial web pages for a web development team. He also went to school part time at Dalhousie University in Halifax and earned his diploma in computer sciences. He attended from 1998 through 2001, and three computer companies funded this education. This was also a COBOL course, and since he was one of the top 25 in the course, he had a guaranteed job waiting when he completed it.

LAND IN SAN PEDRO

A week after he took his exams and received his diploma in Computer Science, Bob came to San Pedro for a holiday and looked at some real estate on the lagoon side that he didn't buy.

"I had a friend from Canada who had bought land here in San Pedro in the year 2000. He told me about it and showed me pictures. That is why I came down, to buy land. I wanted to get away from the snow and retire where it is warm.

"I looked at land in June 2001, and I got offered a job in Chan Chich. While I was here, I met Mrs. Dixie Bowen, Barry Bowen's wife, and I ended up helping them redesign their kitchen at Chan Chich Resort, up in northwestern Belize, in the deep jungle on the Mayan Plaza. I ended up being their chef there for 1½ years.

"To make the change I went back to Canada to quit my job and tidy up business up there, and I moved to San Pedro in Sept 2001 to be the chef at Chan Chich."

After 1½ years Bob left Chan Chich and moved to Belize City in January 2003. He didn't want to live in the bush anymore, and he had met a Belizean girl he wanted to be with. Bob lived in Belize City until January 2004, and during this time he waited for his permanent resident status to be approved.

"To get residency in Belize you have to fulfill certain requirements. It is a long process. You have to be here for a year, and are allowed only two weeks out of the country during this year. After you apply it takes 8 to 15 months to be approved. After applying, you can leave the country if you want to go on a holiday.

"I was living in Belize City with my Belizean girl and I was waiting for my residency to be approved so that I could work here. You need a work permit or resident status to be able to work in Belize. To be able to have my own business, they wanted me to get a work permit. While working for others, the employer is responsible for getting the employee's work permit.

"At Immigration they kept saying, 'Your residency will be approved next month.'" It took 15 months until he received approval in November 2003. During that time Bob helped his girlfriend, who managed the pharmacy at the hospital, to get her own pharmacy business going.

"In January 2004 I went to work as a chef at Captain Morgan's Resort, as the executive chef, and I stayed there until about October. It was okay to work there. I worked for 12 days and then I had two days off. My lady stayed in Belize City."

AGAIN, MANAGING HIS OWN BUSINESS

"In October I moved back to Belize City and started making plans to open a restaurant there. In March 2005 I opened a restaurant called 'Checkers' where I served local food, subs, pizza, fried chicken, burgers, and other items, on Princess Margaret Drive. I rented a house and I built everything myself: electricity, plumbing, had a thatched roof made, and had seats for 40 people. I tried to get a liquor license. I fought with the City Council for eight months to get a restaurant liquor license, which they denied three times, all for non-valid reasons.

"First they were against the thatched roof, but other houses in the area had thatched roofs. Then I applied again three months later. They rejected it and said it was too close to the school, but there was another bar just as near. I applied again. They rejected me again and said that it was too close to a church, but there was another bar right across from the church. Realizing I would never get a license, I closed down the restaurant. The main reason was that the Town Council didn't like that I was competing with Belizeans, with a local guy. They didn't want to let a gringo take business from local business.

"In this country people bribe all the time – for a boat captain's license, to show people land, and other businesses. Most people want around US $500. With the liquor license it was probably the same. Because I could not serve liquor and beer with the food, the restaurant struggled. I closed it up, losing money, and moved back to San Pedro in December 2005. Stress split me and my lady friend apart."

BACK INTO REAL ESTATE

Here in San Pedro, Bob started to work as office manager at a local real estate company, Belize Shores Realty. In May of 2006 he opened his own real estate company, "Coral Beach Realty." His boss at Belize Shores Realty did not know. Bob had a partner coming down from California, who had 33 years' experience in the real estate business, seven in Cancun, Mexico.

Bob has a friend here who is a developer, and he wanted

Bob to advertise his condo units through this real estate company. He knows Bob, and trusts Bob to protect his interests.

"Now that I am opening my own real estate company, my friend's properties will be run through my company. He wants me to ensure good business. If I sell I get a salary; if not I don't get any. This is how it works in real estate."

Ernie has an office in Caye Caulker: a desk, a chair, and two signs. Mickey, the owner of the company, is advertising this part of the business for US $75,000, but it has no listings. The San Pedro part of the business is up for sale for US $175,000. Ernie's partner has a little café in Caye Caulker. "I help with recipes."

In San Pedro there are many open listings from property owners. Whoever sells a property gets the commission. This is the rule. No one gets exclusive rights.

Bob and his new business partner Fred worked on establishing also a mortgage financing company. People can finance their real estate purchases through this company in the US, and Bob has five years of experience in various parts of Belize, so he and his business partner who has extensive experience in real estate are a good match. By June 2006 their new real estate company was off and running. A website for Coral Beach Realty was built in the US and the company's name was registered in Belmopan, the capital of Belize.

"My future plans include making my company a success, building myself a house on my land in San Pedro and enjoying life."

NO CAPITAL GAIN TAX

On the northern part of the island, the Columbians built an airport 20 years ago to land and refuel their planes. Now a shrimp farm uses it. Sueño del Mar is building a huge condo timeshare division with 500 units directly by the airport, 12 miles up north on the northern island, in an area up there called Rocky Point. Anybody could fly in and out secretly at night.

"Between San Pedro Town and San Pedro North there are residential lots, and resorts are being built. They could extend the airstrip up north to make it large enough for international flights. Maybe three, five, ten years from now, you may fly out of the United States directly to San Pedro. This part up north is going to turn into a hotel zone just like in Cancun. Land prices will double in two or three years.

"In the Habaneros area, a 100-foot beach lot was US $120,000; the same now costs $300,000. Second row lots were $25,000 three years ago, but now they sell for $120,000. Put money into land, sit on it, then sell it at a profit. Property value goes up so fast. Some people have cash. If you make capital gains you do not have to pay tax on it here. When you sell, the banks don't release the information to the US Tax Department, so you can gain capital without paying tax on it.

"I recommend, if you have money to invest, instead of getting 5 or 6% on your capital, why not invest down here and get 25 to 35% annually? And another thing, although prices seem high, if you compare them to the rest of the Caribbean, they are still only half. That is why Belize is booming now. Gringos invaded the other Caribbean islands in the '50s and '60s. Belize was discovered only in the last 10 years. Land prices are rising; they will catch up with other Caribbean islands."

Bob realized his dream and founded his own real estate company in San Pedro together with a partner, a US expatriate, and their business is flourishing with the beautiful land and houses they have for sale to investors and new home owners.

PART III, CHAPTER 22: FRACTUAL OWNERSHIP MARKETING AND MANAGEMENT

Generations ago Lee's family came from France and Germany to the US in the 1840s, to the Philadelphia area in Pennsylvania. Originally they were Huguenots.

Lee Albert was born and raised in Kansas, where he earned a Bachelor's degree in psychology and physiology. He taught and coached wrestling at Garden City Kansas High School for five years. Then he went into the sporting goods business and owned a sporting goods store in Kansas.

Later Lee moved to Colorado and went into the land business, where he has been ever since, for 25 years.

DREAM OF THE SEA

Lee is the sales manager and one of the investors of Sueño del Mar, which means "Dream of the Sea".

The company develops land, sells home sites and built a beautiful golf course in Colorado. They sell second home property and recreational property. Land Properties Inc. is the name of their company and this is their 25th year in business. Lee has been with them for over 15 years.

"We have over 6,000 owners who have fulfilled their mountain dream. A lot of them were asking us, 'Now that you helped us to fulfill our mountain dream, can you come and fulfill our ocean dream?' So we started looking and that is how we found Belize. What impressed us most about Belize was that everyone spoke English, all of the paperwork was in English and Stewart Title Company (one of the biggest firms in the US if not the biggest) guarantees deeds and titles.

"The people here are very very friendly. There are direct flights from the US into Belize. A huge part of the US population can get here in less than 2 hours. The accessibility was very good. We felt comfortable purchasing land to fulfill the dreams of our other members. We purchased property here in fall 2004, a little over 10 acres."

Lee's children are grown; he now has grandchildren. His wife comes for one week every month. She likes Belize, but she stays close to the grandchildren in the States. "The kids and grandkids like to come down here to see Grandpa too, to dive, and fish, and snorkel." Lee tries to go back to the US once every six months.

"Obviously it is a challenge to be out of the country. You have to learn their ways. You have to learn about the people, who to contact, who to use. It has been a fun experience.

"Certainly there are things you have to overcome, but it is not really more than in the US. One has to know what the officials want, what they need. If you figure that out, and how they want it, then it is not so difficult. Pay someone to help you to guide you the right direction. They are business people. There are several Belizeans who know the system. Ask around for who is the best person to help you."

Sueño del Mar is a private membership club that basically

allows people to purchase the amount of time they prefer in paradise. Two months is the minimum. People buy two months, four months, six months, a year. If a person owns January and February in an apartment, for example, he owns it until he sells it or someone inherits it after his death. Beachfront property sells for more than property farther from the beach.

"One of the reasons we started this project was to help our existing customers. This system is brand new to us. I think it is very interesting. Over 50% of the new owners say that we have exceeded their expectations in beauty, quality and service. Either they are upgrading or purchasing more time or sending us referrals who are purchasing. They started to live there as of the 1st of January 2006, and the first 12 units are already finished and filled.

"There is a bar, a restaurant, a swimming pool, a dive shop, sea kayaks, and sailboats. They provide shuttle service to town three times a day, at 8:30, 12:30 and 4:30. The resort is situated about 13 miles north of San Pedro.

"The company researched carefully and devised a scheme called fractional ownership, which is a win-win situation. It is fun to see how the people become friends. The people who were here in January and February have already had a reunion in the US. After a week or two, they speak of going back to their home. It is fun to hear them to refer to Sueño del Mar that way, to watch them experience diving, catching fish. I was watching a customer catching a seven-foot shark, and we were excited about it."

When Lee's grandson was here, he earned his certification

to dive. He said that was "really cool," that he was "really happy." Lee and his employees enjoy hearing their customers' stories about diving and cave tubing, which is guided cave exploration by rubber inner tube. Since most of the customers have never been to Belize, everything is a new experience to them.

"Some of the customers have never seen the condos they have bought. Obviously most come down to see it and purchase it down here. They have done business with us before; they have confidence that we do what we say.

"My typical day, I get up, take a long walk on the beach, work out, work all day, go home in the evening and do it all again the next day. On the weekends we try to get up to the resort and enjoy some of the amenities our customers enjoy every day. It is a private club; there is no one you don't know. You can share it with a partner, for example one month each. Especially younger people do that. They don't have so many holidays. But most of the owners are of retirement age.

"There will be a total of 72 units in 16 buildings, with a lot of open space on 10 acres. A little town center, a little grocery store, a conference center. There are two- and three-bedroom units and in addition we'll build six motel units, so there is extra space for members' guests and friends.

"There are nine people working in the office in San Pedro: management, support staff and sales staff. Most interested people just come to us. The units are mainly for use and enjoyment, not so much for investment, but you can use it and sell it later and make money. Our main objective is for use and enjoyment. That it is going up in value is just a bonus."

Like all businesses on Ambergris Caye, it is tourist oriented. In 2 to 3 years, Lee expects everything to be finished and sold out. Then they will begin building the second unit on Ambergris Caye. (Information on Sueño del Mar at helga.peham@chello.at.)

"My focus is sales. We are marketing all over, mainly in the US. We work trade shows, newspapers and magazines, TV – we had a big television show that aired in Denver – and of course the Internet. People are looking for beach property mainly through our website. The biggest source of clients comes from our customer database. They use and enjoy it; they want to use it with their friends and family and send referrals. If they have a fun experience they want to share it.

"I own up there too and I live in San Pedro. I retired twice and they talked me into coming back twice. First in Colorado, to take over a project five years ago, and then, 'You've got to do a little time to Belize,' and here I am.

"It all boils down to customer service, to customer satisfaction. And that stands true wherever we are. Quality product, good meals, anticipate the customers' needs, walk in the door, have fun. You don't have to repair anything. We anticipate their needs and take care of them. We train the people. The resort manager is from the US. We deliver quality meals, drinks, ask them if we can help them. Insist on it, the staff figures it out quickly. Belizean people are happy and friendly, and we can help them service. That is a good start if the staff has already that part.

"I'll do it as long as I am having a good time, as long as I am enjoying it. When it becomes work, I will stop."

PART III, CHAPTER 23: A UKRAINIAN-AMERICAN

LEAVING UKRAINE

Vladimir Shkavritko, called Walter, was born in Ukraine. His family lived there until he was nine years old. His father was an electrician and his mother was a stay-at-home mom.

Walter's grandparents had a condo, but they were persecuted for being Pentecostal Christians. In 1989 Ukraine was still a Communist country. There was a program offered for people persecuted for their faith. They were offered visas to leave the country.

Walter's parents decided to go. They left everything behind and only took a suitcase with the bare necessities. The whole family went together: Walter's grandparents, his uncles, his aunt and her family, his parents and his sister, even his great-grandfather. His cousins joined them later.

AUSTRIA – A BELOVED STOPOVER

They passed through Czechoslovakia and arrived in Vienna, Austria, their first station. They stayed there for two or three weeks. Walter's father didn't want to leave. For Walter it was amazing to see what they sold in Austria. "I liked it. The stores had everything." At the markets they could pick up the leftover fruit at the end of the day. They lived in an apartment building where a series of rooms were available for refugees, and they received some money to buy food. Walter wants to go back to Austria some day.

"I was just amazed, everything was new over there. The cities in Austria were clean, the shops full of stuff, they sold lovely ice cream. Everything was well organized."

They stayed two or three weeks in Austria. A lot of paperwork had to be done before they could go to the US, their final destination, where both the states and the churches sponsored families. This all took three to six months.

ITALY

They were brought from Austria to Italy where more paperwork was done. Italy was nice too, but they preferred Austria. They went to Rome and found lodging just outside the city, on one of Rome's hills, in the Hotel Santa Barbara. In Italy they were provided with housing and food: breakfast, lunch and dinner, mostly pasta, sometimes with cheese.

USA

The family was sponsored by a church in Rochester, New York. They flew from Rome to JFK in New York City and then to Rochester. An uncle and his two children had arrived two months before them.

When the USSR fell in 1991, the program fell. It had been a program for those who were persecuted because of their religion. Some Jewish people had also left the USSR, with the choice of going either to Israel or the US.

GETTING SETTLED IN THE NEW WORLD

His parents had US $1000 and a suitcase, and they couldn't speak English. Walter started school in the second grade in Rochester, but he didn't speak English and could only count from one to five. The people in the church looked after them, gave them free furniture, and helped them settle.

His parents and his grandmother went to school for a year and learned English while his great-grandfather stayed at home and looked after the children.

After one year, in 1990, Walter's mother got a job in a plastic company that made bottles. She worked the night shift, and during the day she continued going to school to learn English. His father was still going to school.

Walter has since been to most of the states in the US.

SCHOOL, COLLEGE AND WORK

By the end of third grade, Walter spoke English. It took him about two years to understand and speak English. "In the fifth grade I was so good that I skipped to seventh grade." His sister went to kindergarten and spoke English after one year. His parents were working and going to school. His mother worked in the plastic factory for five years, his grandmother for three years, and his father for one year.

Walter finished high school in 1996. He left after 10th grade, skipping the two final years, and went to college from 1997 through 1998. He attended college in the daytime and worked in the evenings. "I started working at a restaurant when

I was 13, 9 hours a day, Friday night and Saturday night." But that wasn't his first job. When he was 10, he helped his dad to paint houses.

In 1993, Walter's father was working at an Alton machine shop. He learned a specific job and did it repeatedly. Then he went to work in the restaurant, Grinells, as a dishwasher and took Walter with him. That is how Walter came to work in this restaurant, as a busboy. Walter did this work for five years, and several other members of his family were also working in this restaurant. After his mother left the factory, she also became a dishwasher at the restaurant.

Walter was working in the restaurant two days a week, Saturday and Sunday. He earned US $50 a night, or US $100 a week. He saved US $2000 in one year, bought a car for $2000, and sold it for $2800 after a little work on it. He was 14 years old. At the same time he helped his father with odd jobs, and he interpreted for his parents at doctors' offices and anywhere else they needed English.

When Walter was 14 he bought an old car, fixed it and sold it. A year later, at 15, Walter bought and repaired another car. When he was 16, he crashed it. The car was fully insured so he received its value in cash. He continued buying, fixing and selling cars. His next car, when he was still 16, was a Porsche. He was 19 when he started selling new cars.

Walter did construction work when he was 17 and 18, while going to college. He worked during the day at construction sites and went to school at night. Some weeks he worked 80 hours. When school was out he worked 40 hours. He went to Monroe Community College, the third best community college in the US, with 30,000 to 50,000 students. He started in radiology, then switched to computer technology. His cousin did the same, then switched from hardware to software. Walter switched to engineering and did that for a year.

Walter wanted to have something of his own. "I always wanted to grow, I didn't want a ceiling, so the only way to do this is by working for yourself." He didn't graduate from college. "I had lots of credits, but they didn't match." In 1998 he left college.

VISITS TO THE UKRAINE

In 1994 Walter, his dad and his grandmother went to visit Ukraine. A year later his great-grandfather returned to Ukraine from the US. Everything had changed. Ukraine looked "like it had been through a war. Everything was totally different to what I had expected. Even the people changed. They were very poor." Walter stayed at one of his aunt's houses. They also had to gradually pay back the ticket from 1989.

In August 1998 Walter went with his mother to Ukraine again. The country was growing. It had more foreign cars than before, and more foreign products on the shelves, but the people didn't have the money to buy them. Walter and his mother stayed there for a month and a week.

"At first, there was a lot of mafia around, you had to be careful. There was organized crime. Fees for the protection of a store were collected. When you opened a new store you had to pay. If you had a nice car you had to pay for protection, otherwise it disappeared and someone called you and asked you to pay if you wanted your car back.

"But after they had made enough money, the mafia down there was slowly turning into businesspeople. They tried to become legal.

"There used to be a lot of corruption in the police and in government. Now they send undercover police officers to check most of the corrupt policemen out. Those corrupt policemen are fired."

In 2004 Walter returned to Ukraine for 10 days to research the possibilities of his partner and himself building condos at Ivanov on the coast. In 2005 he returned for 1½ months.

AFTER COLLEGE

After Walter had left college he bought a tow truck with a flat bed. Again he bought cars, fixed them and sold them at a higher price. He also delivered cars to other people, which is how he traveled through all the northeastern states in the US.

Walter always wanted to invent something new. "That is what I am still working at." He has always had a lot of good ideas.

Walter opened a detailing shop to fix cars for other people. He also found cars for customers in the price range they wanted, picked them up and delivered them. Walter had this shop, together with a friend, for two years.

One day a pharmacist wanted a two-year-old BMW. Walter found a couple of these cars and showed him the pictures.

Walter picked up the car for him, typed out the contract, got the $40,000 check, cashed it and got his commission from the seller.

The pharmacist was very pleased with the car and the whole transaction. He wanted Walter to work for him in the pharmacy. "I wanted to try everything." He worked mornings for the pharmacist for one year. Then he closed his detailing shop.

BELIZEAN PROJECTS

In 2003, the pharmacist went with his sister on a vacation to San Pedro in Belize, on the island of Ambergris Caye. The same week he bought 10 acres in the south of the island.

He asked Walter if he wanted to go to San Pedro to work for him there. Walter hadn't been to the island yet, or to Belize, so it was hard for him to make a decision. The pharmacist offered him time and money. Walter agreed to the project but he had no US passport. He was stateless. He got a Ukrainian citizenship card. Then he was able to get a passport and a green card, and he could go to Belize.

Walter came here in 2004. He had been to the Bahamas before. "When I came down here, everything was different here." It was a difficult situation for him. He wanted to do business down here but didn't know the country. He stayed at Martha's Hotel for three years before moving to an apartment in the post office building for one year at US $600 plus utilities.

Walter had come to Belize with a plan to do things in a certain timeframe, but everything took much longer than planned.

He had to run around to get things done, gathering all the necessary papers and learning as he went along. This was a frustrating experience. "I had to get used to the way things work down here."

His project was to bring jet skis down to Belize, so he shipped a container from the US. People rent jet skis and go along the coast around the island. But Customs are a huge problem in Belize, and problems must be solved in Belize City. "In the beginning I didn't know which papers they wanted. It was a lot of hassle getting things started."

His business partner intended to start a resort at the south end of San Pedro where he had bought a large acreage. Walter did the research on the resort and got the marina started, but the timing was wrong. San Pedro was working to attract businesspeople, but there was not enough tourist traffic. That wasn't Walter's only problem. "Unfortunately the government gives you a hassle when you try to do anything. First people struggle, survive and make it in the end."

There are a lot of advantages to choosing Belize as a place to live and to invest. Belize is the only English-speaking country in Central America and it is a safe place. British Common Law is used and you don't have to be a resident to own property. Two Belize dollars equal one US dollar, which makes the math simple, and US dollars are accepted everywhere. The weather is nice and warm and Belize has the second largest reef in the world.

Belize is a good place to invest because of the taxes. Property values increase, taxes remain low, and the banks pay good rates. Prices for property have tripled during the past five years. Prices are expected to double within the next five years or so.

Of course this depends on the area. This is especially true for areas that grow fast. The north and south sides of Ambergris Caye are both fast-growing areas. All of Ambergris Caye will grow fast. "In a few years you will pay much more for property."

Walter and his partner have 40 acres on the northern part of the island, at Nella Reed past Basil Jones. They also bought two lots on the lagoon in San Pedro and built a house there. Walter manages the entire project.

TOP QUALITY CONSTRUCTION

Walter took over a construction company and named it Top Quality Construction Ltd. He offers top quality building with a fixed ending date, something that is not usual in Belize. Most building projects take much longer than projected and cost more than agreed. At present he is building a few two-unit condominium buildings in the south. San Pablo is a good area to live in.

Global Real Estate Ltd. is Walter's own company. His business is helping other expatriates with their projects.

He can sell lots and build on them. He has catalogs from stores in Miami. His customers here can choose what they want. They have many more options than are offered in Belize. Here, often the things run out of stock. Walter has containers come from Miami with everything that is needed, all finishing

...ucts and accessories. Granite flooring, ceilings, sinks, cabinets, trim, and tiles are brought in. Walter builds in concrete and wood, and offers various options. He handles all of his customers' construction needs.

INVESTMENT PROPERTIES

He also finds investment properties, subdivides them and resells. He wants to get customers for a lifetime. Walter invests in tools, finds the best people for the jobs and trains his people. In 2004 he came for one year. He comes and goes. The island of Ambergris Caye has a lot of potential and so have Placencia and Cayo. He subcontracted the jet ski business.

Walter was not in Belize during the low season in 2005. In 2006 he stayed in Belize. He is building houses and condominiums and fulfilling other contracts. He has 15 people working for him: masons, carpenters, electricians, plumbers, laborers. He wants to make sure that people do not pay too much for their houses. Walter offers quality plus the assurance that the house is built as close to the agreed timeframe as possible.

Walter also wants to open a customer service company to reduce expenses for companies anywhere in the world. He is doing some research into this project now. He needs Internet and telephone, and he can employ people in Belize to form his new company. He is open to new business opportunities. "I don't like to limit myself to a few things only. I like variety." His jet skis are the only ones in Belize, and this country offers a variety of water sports. He is also looking into offering hurricane insurance on homes.

Investors are needed here. He is opening a mortgage company. Here interest rates are up to 15%, but in the US they are only 6%. Walter wants to offer mortgages at only 8 or 9%, 6% for the investor and 2% for him. Everybody wins. This is a safe investment because properties increase in value. People who need mortgages will finance them. This is more of an international arrangement, not a Belizean one. Lots are 50 x 100 feet and 75 x 100 feet.

The government plans a city on the northern part of the island. If the Belizeans build a road to the end of the island, the Mexicans will build a bridge so vehicles can come in directly from Mexico. The mayor of San Pedro is one of Walter's best friends in Belize. It will take 10 years for all these developments to be done.

There are a couple of roads needed in the south of the island. They must be built with private money. The people living there want these roads, so the mayor can use their money to finance this project.

There has been some development on the northern part of the island, and some bigger plans are on hold. In a new town up north they had plans for a casino, but this is also on hold. Taxes on casinos are much lower in Belize. The plan is to fly people in to enjoy themselves in the casino.

Walter believes there is much more stress in the US, but on the other hand, things happen faster in the US. The Belizean lifestyle is much more relaxed, with a slower pace that might frustrate an impatient expat. Breaking through in Belize is the hardest thing to accomplish. Walter is prepared to show people who want to settle and do business in Belize how to break through.

ISLAND CHARACTERS

Paysano was "the old man with the stick." He once was a wealthy man who came from the Middle East, some say Lebanon. He got into drugs and the government confiscated his belongings. He then was homeless and slept on the docks. At any time he could go into the shops to get something he needed and he never had to pay. He was a sort of a legend of the island. He died in December 2005.

Leonardo DiCaprio was in the Caye Esponte Island Resort. The staunch environmentalist DiCaprio is planning to turn Blackadore Caye, off the coast of Belize, into an eco-friendly resort. The famous actor plans to build this environmentally-conscious retreat of a few exclusive villas with private pools and terraces on the sandy beach, while respecting the island's wildlife in the tropical surroundings. It is costing him millions of dollars to create.

A lot of movie stars come to the island. Most you will not recognize. Robert De Niro came to Belize for fishing. Madonna made a video down here and sang her song 'La Isla Bonita' on San Pedro. In an interview with the New York Times, she called the song, "A tribute to the beauty and mystery of Latin American people."

Realize your dream

Looking for a House?

Fido's Pier

Looking at the World from the other Side

Come diving in San Pedro

Dusk

PART IV: ARTISTS

PART IV, CHAPTER 24:
BELIZEAN ARTS

"The first time I heard about Belize, I was living in Jordan and working for Jordan Airlines. I traveled constantly, and I heard from coworkers that Belize was a fantastic place to move to. My partner, a pilot, had lived in Belize in the '70s. Of all the countries he knew, he thought Belize was the most perfect place to live."

Lindsey Hackstone was born in West Kirby, England, and grew up in Guildford in South England. All her life, Lindsey's parents encouraged her and her sister to participate in and enjoy arts. Her father took her to the ballet, art galleries and the opera. At an early age, she knew that art would play an important part in her life.

When she had nearly completed her studies, she left England to explore the world. She lived in Forestville in northern California for four years, then moved to South Korea to work as a ticket agent for British Airways.

FIRST IMPRESSIONS

"I first came here in 1983 on a vacation. The first thing that struck me on the flight from International Airport to San Pedro was the different colors in the sea – so many islands and so many different colors in the water. I loved it. I never knew at that time that I would end up living here. It seemed like it was cut off from the rest of the world, and it was in those days."

'Soaking up the color in San Pedro' was the title of an article about Lindsey that appeared in the San Pedro Sun in July 2001.

"A few years later I ended up living in El Salvador in 1988. My husband, an American, was a pilot for Taca. El Salvador is beautiful, and he was stationed there, but the war hadn't ended. We bought a piece of land in San Pedro, moved here, and built a house. But soon after, we split up.

"The art gallery was a hobby. Belize art is interesting art. I love native art. Therefore I opened a gallery. I stayed here, had three children, and turned my hobby into a moneymaking business. I opened the gallery about 1989. It was financially a struggle for many years and really I think I love art, I loved what I was doing and in the end made a profit. This is more a love than a business. It was a struggle, and for years I was broke, a good experience. I now know what it is to be poor. I know now how to up from being down."

Through traveling Lindsey has developed to a taste for ethnic art.

"The Belizean Art movement was very limited at that time, three or four artists. Louis Belisle, Benjamin Nicholas, Pen Cayetano, and Nelson Young and Philip Lewis. These guys were painting because they wanted to paint.

"Pen Cayetano had his own gallery on the main street in Dangriga. He is known for his cultural paintings, often depicting Garifuna rituals like dugu and others. He is also known for his music. I was really impressed in his little place, painting when hardly anyone was coming to Dangriga. He has evolved incredibly. He is a fabulous artist. You feel the heat and the rhythm. He is Garifuna. Now he lives in Germany.

"Nelson Young at that time was living in Caye Caulker in a tree. He is a Rastafarian. He had recently come back from Europe and Sweden. He showed his brethren, his brothers, how to live the clean and healthy life with little money. You don't have to live in dirt if you are poor. He would bake fresh bread and cook meals over an open fire. His old paintings would be hanging from the limbs of the tree. He made sure he did not nail. He used wood, but no nails for his house in the tree. His home was a piece of art, a sort of installation, a very creative achievement. Later the Town Board cut it down. The food he prepared was great. Those days most of his art was very native. Now he has moved into abstract. Today he still lives in Belize on the cayes and moves around. I love his art; he paints for the right reasons.

"Louis Belisle passed away. His paintings still can be seen in the Fort George Hotel. Nicolas lives in Dangriga. He paints historical paintings and cultural paintings about Garifuna.

"The art scene has changed due to the Image Factory that opened ten years ago. And also galleries like my own interest the travelers in local art. Now there are art customers. There is an opportunity that was not here before. There are lots of artists now.

"People who buy art focus mainly on paintings, but they also buy prints of paintings. Jewelry made by local artisans is popular. We also have some ceramics and we have a huge number of canvases, a large number of paintings. Our main goal is to support local artists; they get the greater part of the sale, because we work closely with artists. I paint. We make sure the artist gets more than the gallery."

(Interested in Belizean arts and crafts? Then send an email to helga.peham@chello.at.)

LINDSEY'S ADVICE

"If I was to describe my business, basically it is an anthropological field. You cannot talk to the artists about marketing like we do in the West. People have their ways here. I have enjoyed this. During my living in Belize, I have enjoyed their different approach. I find that very fascinating.

"If people do want to bring their ways down here they will not be happy here, because it is a different culture. Expecting things to be like where you are from is not realistic; you have to know that you are entering another culture. Sometimes amazing things happen here. People are amazingly helpful. To enjoy the differences is important, not to fight them.

"Business opportunities are mainly in tourism. People who want to invest in building resorts and hotels can do very well here. Also, investing in real estate is a good business decision. The cayes are the hottest area, making money: San Pedro, Caye Caulker and Placencia. There are still opportunities here. You have to go through the same paperwork as in other countries, but that is normal and takes time. Some people have made money buying condos. The prices are rising. From Cancun to

Chetumal prices rise. This will happen here as well.

"The fact that Belize is close to the US, just two hours away by plane, is favorable. It is still affordable to buy beachfront properties. They are not as expensive as in Playa del Carmen.

"Also people speak English. You can speak with locals when you come down here. I think Belize only recently has been discovered – one of the last frontiers left. No big corporations or large businesses have taken over. The people are very charming.

"People who dare more are more successful here. Degrees may not help. You just have to make things happen."

PART IV, CHAPTER 25: A CARIBBEAN MUSICIAN

"My childhood ranged from deep sorrow to happiness, everything in extremes, but that was appropriate. I grew up in a country where everything was in the extreme. Always celebration, many musicians, a lot of artists, playing a lot of instruments. There were numerous ethnic groups: Hindus, Chinese, Whites, Negroes, Muslims, Catholics, all types of religions, a real cultural mix. The generation I grew up in changed many of the old standards. The way of thinking changed in that generation when I grew up. I saw many social changes, and political changes from colonialism to independence."

The artist and musician Lennard M. Bonardy, called Lenny, was born in Port of Spain, in Trinidad, in the Republic of Trinidad and Tobago.

OIL BUSINESS

Trinidad and Tobago experienced an oil boom in the '60s and the '70s when Lenny was young, somewhat similar to what Belize is going through now. The oil boom in Trinidad and Tobago brought a lot of investment into infrastructure development and employment, technical training, and an improvement in the standard of living. There was a very active tourist industry in addition to the oil business. A great deal of money began circulating. International corruption, guns, drugs, and political scandals all came with this development.

REVOLUTIONARY IDEAS AND SOCIAL CHANGES

"These changes made me a very revolutionary individual. Not only in arms but also in ideas as a social worker and a teacher."

Lenny struck a balance with the school he worked in. It was a relief for him to stop singing 'God Save the Queen.' Lenny was affected by growing up among people who contributed a lot to deep political changes – intellectual, musical, financial, standards of living, education, human rights and workers' rights. Within the society they needed to coexist.

Lenny's generation could not change the old patterns of thinking. Colonial people had a way of holding on in desperation and hoping for a better life for the population in Trinidad. In contrast, "We got up and we made the changes."

A BRITISH COLONY

Trinidad was a British colony. It gained its independence in 1960 and became a part of the British Commonwealth. Lenny knew the face of colonialism and the political games, that they were manipulated.

EDUCATION

Lenny remembers well his first years of education. He spent five years in primary school and became a very good painter. Commercial art became his specialty.

"We attended primary, starting with kindergarten, learned the ABCs with a very strict old lady with a big cane. We had to learn the ABCs whether we liked it or not. I went full term, seven, eight years in primary school. I started school at five years. From primary school I got into intermediate secondary school because of my ability to draw and paint. There I learned to use what I know. I knew to draw and paint. They taught me in school how to use this in a commercial way."

Lenny grew up with a lot of extremes.

"My father died when I was only six years old and there were nine of us, four girls and five boys. That was the tragedy. We went to school and we had a normal life. We were all adopted. We all reunited with our mother after a few months. With me it was about a year. We all ran away and came back to mother and grandmother. My mother was a beautiful young Chinese lady. She worked hard. She did boarding. People came to eat lunch. She did sewing, made clothes for people on a machine, and draperies for houses. Plus she went downtown to sell imported fabrics for other people."

INTERCULTURAL MARRIAGES

Lenny's father was an African descendant from Martinique. His mother's father was from Hong Kong and his mother's mother was a Chinese born in Trinidad. They are called Creole Chinese. There are a lot of Chinese people in Trinidad.

"In Trinidad there are a great number of intercultural marriages, interracial marriages and relationships, so we get a lot of half-breeds. The children of these people change the society, this half-breed nation, we change the society. Traditional marriage doesn't interest me. A spiritual relationship would appeal more to me, based more on what we see."

MUSIC

Trinidadians love music. They are musically and artistically oriented people. Young people can learn to play music in classes which are sponsored by the government. Lenny learned about music while going to technical school doing commercial art.

"We were playing music at home singing, beating drums. We grew up among musicians."

A music teacher from Trinidad trained them in theory, up to grade five or grade six, up to the point of transposing a melody from one key to another. Lenny and the young people in his class learned songs and listened to other groups.

"We had a family band: my brother played keyboard, I played lead guitar, my other brother played bass guitar, and a cousin beat the drums. It was a real family band."

They had just a short period when they played together. Later on they played in different groups in Trinidad and Tobago. Very early, after playing for about three years as an amateur, Lenny got the opportunity to play with a big group.

"I never saw so many musicians living together in one place like in Trinidad and Tobago."

PLAYING IN A BAND

After school Lenny worked for six months and did some technical drawing for a while in the mid-seventies.

"There were designers, display designers, arranging shop fronts. I worked for them for a couple of months."

Soon, Lenny stopped working for other people and became self-employed doing commercial art, freelance and outdoor advertising, and he has remained self-employed ever since.

In 1977 Lenny left Trinidad and spent about three years in Venezuela, playing music and painting.

"I went there with a whole band from Trinidad, a whole group of people: drums, keyboards, guitars and vocals. Seven of us went to Venezuela, including my brother. We played a lot of music there. We went right up to the top. We had good equipment and a good job at the Intercontinental Hotel in Porto Daz in Venezuela. That was in 1980."

Lenny really can't say why he left Venezuela, but he returned to Trinidad and stayed there for about 18 years, until 1998. "During that period I had a wife and five children."

Lenny stopped playing music professionally. He only did commercial art. They lived in different places. They had to rent and move and rent and move. "We had to move due to financial reasons, with five children and a wife and I was self-employed."

MARRIAGE AND FAMILY LIFE

How did Lenny meet his wife? Both their families grew up in the ghetto in downtown Trinidad. Lenny knew her from his childhood. She was always clean, and helpful and happy. They decided to get married because she wanted to live with him and he wanted to live with her. "I still love her. Love does not change."

She knew how much he gave up to have a family with her. She knew that Lenny always wanted to go back to playing music. "All my friends knew that I was a musician. I gave up playing music. My children didn't see me as a musician."

In 1998, he left his family to come to Belize with the full consent of his wife and his children.

BELIZE – A LIFE FULL OF MUSIC AND ART

"That was the last thing my mother did for me before she died.

"She was moving to Belize to live. She had been living in New York for about 10, 15, maybe 20 years. She and my stepfather immigrated to the US. My oldest brother was a Vietnam veteran so my mother got the papers for the US quickly.

"She bought tickets for my brother and me to come to live in Belize. She wanted us to experience its culture and progress. Our mother knew we were deeply attached to music and made it possible with encouragement and finances.

"She went from Trinidad to New York. That was the last time I saw her. It was at the airport. When she went to New York we went to Belize. From New York she was making preparations to come to Belize to make some investments. Unfortunately she died in New York from a heart condition. That was five years ago, in the year 2000.

"So here I am. After living a married life for 25 years, my presumed destiny is to be a musician. The children have all grown up and are working. My family is OK."

When Lenny first moved to Belize it was the beginning of a new educational experience for him.

"I think I know what I experienced. I have the time to reflect on knowledge and truth, mystical knowledge. An example: traditional society believes that god is outside of us. But in reality god is inside, it is not outside of us at all."

Lenny's sister Cheryl lives in Crooked Tree. She married a Belizean. Lenny spent about three months in Crooked Tree. After that, he and his brother went to Caye Caulker.

"I spent a year without shoes, without watch, without motorcars. I financed my life through playing music and painting. We did commercial paintings in places like Rasta Pasta, Oceanside Bar, interior decoration. My brother plays keyboards and I play bass guitar."

Outside of tourist season there is no work for him as an entertainer or as a commercial artist, he explains, because during the offseason, the hurricane season, businesses close down.

Lenny and his brother left Caye Caulker in 2001 and went to San Ignacio, where they spent two to three months at a place before moving on to Placencia and then to Punta Gorda in southern Belize.

Every year they come to San Pedro, because they believe this is the best place to be when the season is right, from December to April or May. San Pedro is their base and Lenny operates most of the time from San Pedro. He can get both music work and commercial art work here. He usually spends some months in Caye Caulker, especially at Christmas time.

Caye Caulker is affordable and clean. They stay at Albert's Mini Mart. In the offseason (June through November) they like to be in Caye Caulker. They play music at the Oceanside, or the Barrier Reef Bar and Papaya's Beach Bar. In San Pedro they play music at Captain Morgan's Resort in the north, or Barefoot Iguana, or El Divino. They have also played at the Blue Water Grill.

Lenny and his brother have plans for the future. They hope to assemble a five-piece orchestra, to go together to Cancun, Costa Rica or Panama.

"This moment here is the important moment. What is good for me is that I come to realize something: life always prepares for the next moment, for the next step. Here, the value of the experiences comes shining through really clearly. This makes me reach a conclusion with greater clarity."

Lenny also makes sculptures and paints on canvas. "We divide our time between painting and playing music. A very creative, very expressive life, a deep and wide and good and perfect life."

"There are three of us, our group of artists. Frank Cadel, he is a Belizean and does sculptures and woodcraft, as do my brother and myself." When they have enough work to display, they will hold an exhibition.

"All my experiences from before prepared me for this moment and the opportunities that exist: the exhibition, playing music and painting and living, music and art and good public relations. This is what it is based on. I am equipped for it so it is happening. Interaction with people on all different levels is important in art because a musician is always exposed to a wide variety of people as an entertainer. All are flavored by this interaction."

There is less work for Lenny in Belize than in Trinidad, but he feels that he gains more self-realization.

"I implemented concepts in my life, fine tuning, public relations, to the point where we have values and understanding and enough faith and strength to face the unknown, to face the future."

SPIRITUALITY AND SPIRITS

Lenny does not hear very well. He inherited that from his mother, who used a hearing aid. This disability started to affect him and his brother when they were between 15 and 17. When he started playing music they used a lot of amplification, which caused their hearing to deteriorate faster.

"It is a paradox having to live life as a musician with a hearing impediment. Hearing comes from outside and it goes inside. When you don't hear very well you hear inside: you listen to the voice of your spirit, mind and your imagination. The songs within the imagination, like harmonies and going along, just exist. It is the mind's ear. That internal part which we do not pay attention to becomes attuned to the internal spirit when you do not hear the outside. Beethoven wrote his best music when he was deaf. When he didn't hear outside, he heard inside and wrote it down. Every day I keep practicing reading sheet music.

"It is important to know that due to my internal hearing I became acquainted with the awareness that is the spark of my life. That awareness is not Lenny, that awareness is the spirit that occupies the body. I have evolved from religion into spirituality. I grew up in Christianity, know about Hinduism, Baptists, Evangelism. Living in a multiracial society exposes you to all the different concepts. I had experiences with all of them. I evolved from them to what I am now. I am a more balanced human being now, physically and spiritually."

WHAT TO OBSERVE WHEN COMING TO BELIZE

"As a general rule, people should be aware of their own likes and dislikes and accept that other people, just like they do, have their likes and dislikes, too. What I would recommend is not something external, but something internal. Something like, if you come to Belize come with your happiness, don't come looking for it here. If you find it, you find it in yourself, right inside. People travel all over the world looking for happiness and they don't know that it is right inside."

PART IV, CHAPTER 26:
AN ARTIST'S LIFE

OFF TO SAN PEDRO

Jim Diehl has been an artist for over 40 years. He moved to San Pedro in 1994. He is a disabled veteran who served in the US Navy from 1965 to 1971. "I was in and out of Vietnam on a ship. In 1971, I left the Navy. I was injured after an epileptic attack. Since then I have been on a pension."

Since his youth he has been an artist. Born in Butler, Pennsylvania, he sold his first oil painting at the age of 16. Art in its variations are Jim's passion. He has worked in water colors, oil paintings, acrylics, commercial art, jewelry design and jewelry making. He has designed ladies' clothes, sculpted, and worked in wood, clay, and stone.

Jim has also spent his life studying another passion, applied physics. "We have a physics think tank here in San Pedro, Panton-Diehl Applied Physics. Its catch phrase is 'Science for Belize.' Winston F. Panton and I are the members. Other people are coming in and out. We talk often on the phone and are together several days a week. I am on the hurricane board here, the San Pedro Emergency Committee/National Emergency Management Organization (SPEC/NEMO). I am the EOC Supervisor/Deputy Operations Officer."

TIRED OF COLD WINTERS

A friend of Jim's was working in Belize for an oil company and told him about the country. Jim felt he had to check it out. This was in April 1994. He came to San Pedro and bought a beach lot. Jim rents an apartment on the beach, close to downtown.

His friend had told Jim to go to the Belize International Airport, and then to the island via Tropic Air. San Pedro was a lot smaller – only 2200 to 3000 people lived there at that time. Now there are about 12,000 residents, not counting tourists.

"I was just tired of the winters. I was in the tropics a lot when I was in the military. When the opportunity presented itself I tried it out to see if I would like it. San Pedro was still pretty quiet. Within a couple of months I was working with the police department through the San Pedro Citizens' Crime Committee of the Chamber of Commerce. I also volunteered for the fire department. In 1998 I participated in SPEC/NEMO during Hurricane Mitch."

PRODUCING ARTWORK IN SAN PEDRO

Jim started producing artwork in San Pedro in 1995. He created a collection of model boats using coconut shells and seed pods, along with mahogany and other woods. Some of them he sold and some he still has. Jim has always changed formats as an artist. Whenever he gets bored with a format, usually after five to seven years, he changes.

"I consider myself a Da Vinci artist. I invent. I create. My company name is 'The Original Stone Art of Belize.'"

Jim creates sculptures, does coral reef scenes, and sells to tourists and locals. His first studio was destroyed by Hurricane Keith, so now he is in his second studio. A variety of galleries are his clients. He sold three pieces to a reputable gallery in New York City for US $2,200. They resold them for US $10,000. Some hotels in San Pedro and Belize City buy them for decoration. Word of mouth is his way of advertising. Jim's prices range between Blz $5 and Blz $2000. A medium-sized work of his stone art might typically sell for Blz $80, and a large stone relief might sell for Blz $1900.

"It is a struggle now, because tourism business is low.

"The material I am using is reconstituted limestone. I do my pictures in plasticine. Then I make a latex and fiberglass mold. After that I cast the reconstituted limestone, then I colorize it and recreate the stone. The colorization process I use was originally developed 400 years ago in Italy. Architects have been using it during all these centuries. I've used it in my art for 35 or 40 years. I do my sculpting, my mold making, here in this studio. As an artist you just learn things through observation. I have seen how it is done in the States and here. I do original work, a lot of coral reef themes. It is very strong material. I do commercial things also: ashtrays, Belikin beer, Lighthouse beer."

In San Pedro one usually socializes in one of the bars. Sometimes business is done in the bars, especially at Happy Hour. Jim likes to go to Crazy Canuck's Beach Bar, to the

Exotic Key Resort south of Belize Yacht Club, and to BC's Beach Bar and Barbecue. "There you get the best barbecue in the country, just south of Sunbreeze Hotel, on the beach, over a rum and Coke."

TRAVELING BELIZE

Jim has traveled throughout Belize, partially because of his involvement in the hurricane watch. "Prime Minister Musa said we had the best NEMO team in Belize. He sent us to Placencia to rebuild after Hurricane Iris struck in 2001."

BE WELL PREPARED

"If you want to move to the country, check it out and be prepared properly.

"Also, there should be more small manufacturing here in Belize that would create jobs for locals.

"For people who want to move here it would be best to come and create a situation and create jobs. San Pedro is not cheap. It is rather expensive to live here, compared to some other parts of the country. But it is still cheaper than the US if you want to do so."

PART IV, CHAPTER 27:
CORNELIUS MAGNUS HARRELL

INTRODUCTION

I met Dr. Cornelius Magnus Harrell, a sculptor and painter, by chance. He is known as Dr. Magnus, and his friends and acquaintances call him Magnus. A mutual friend introduced me to Magnus when he went to the Island of San Pedro for a few days and I happened to be there too. It was 9 pm when I entered the room of a small business office. Dr. Magnus was sitting there with his white cap, and had the air of a flight captain. He welcomed me kindly and graciously.

His face was open and his eyes vivid. Though he could be mistaken for a man 20 years younger, his birth certificate confirms 83 years of age. His dark complexion contrasts with a short white beard. He was ready to tell me his story. We met several times and we talked for hours on the terrace facing the blue sea. Not far away, the Barrier Reef can be seen – the second largest reef in the world. He sipped his Belizean rum from time to time while smoking, and he often got emotional about the stories he shared.

Magnus is a painter, a sculptor, an industrial designer, and a philanthropist. We became friends.

10,000 PAINTINGS

Magnus has 10,000 beautiful framed paintings stored in Florida. These paintings are the result of very hard work after losing his wife, his brother and his mother in the span of a few years. The pain was so strong that he often had the feeling he could not go on. All the sorrow and the energy in him poured into his creative power and in turn into his paintings. He has previously sold his paintings in his galleries in Buffalo, Atlanta, Florida and Belize City, and he has auctioned them off at Sotheby's and Christie's in New York.

DRAWINGS AND PAINTINGS

Magnus has produced 3,500 drawings since August 2004 as well as 300 paintings and collages. Many of his works are sold here in Belize. Magnus taught Prime Minister Musa's son the preliminaries of modern painting, and he introduced modern painting to Belize some decades ago, hitherto unknown to the country. He created the vibrant paintings which illustrate this book.

2006 A NEW PERSPECTIVE OF LIFE AND PAINTING

During 2006, Magnus was restructuring his life. With the sad and hard years gone, there was also a reduction in the output of his work. Current works embrace Minimalism. It penetrates the sketches that he produces every morning after getting up, before the sun shines into his room.

Magnus also designs pottery, loves to design buildings, and knows the arts of welding, industrial design, and clothing and furniture design. This industrial designer, painter and sculptor has produced a massive amount of art during his lifetime. Magnus has left mourning his lost family behind him. Working creatively is a real need for him and therefore he restructures his surroundings accordingly. He really loves to work and needs the surroundings in his life that further his creativity, his style of life, and his well-being through art, design and architecture.

(See Magnus's artwork at the end of this chapter and contact helga.peham@chello.at if you are interested in Magnus's colorful studies in giclee copies)

CHILDHOOD AND YOUTH

Dr. Cornelius Magnus Harrell was born on October 10, 1924, in Toronto, Canada. Magnus's beloved stepfather raised him, together with his dearly loved mother. His real father, born on a Canadian Indian reservation, had immigrated to the US. He returned to Canada later, where Magnus was born. He died when Magnus was just a few years old. His family championed his talent in art and helped provide him with commensurate education. His parents always told him, "Every year, you must learn something new," and he has done so throughout his life.

Magnus has both Canadian and US citizenship. Magnus and his parents went to the United States when Magnus was six years old. They moved to Buffalo, New York. During school breaks, Magnus went to Canada and learned a great deal from his uncles, who were self-employed and prosperous men.

He was never interested in baseball. He doesn't like teamwork or team sports. He prefers one-on-one sports. He loves boxing. As a child he used to run home from school from the bullies and the bad guys.

His mother taught him how to dance, and he went to Argentina to learn the tango. He also went to study art. Magnus refers to himself as an impressionist.

Magnus's father was an architect and builder who was not allowed to work in his profession because he was of African descent. In 1941, he was appointed Professor of Mathematics at the University of Buffalo.

Magnus's mother was a librarian at the university who worked her way up to become the Dean of Library Science at the University of Buffalo. She was biracial – half Indian.

During his school days, Magnus became acquainted with Frank Lloyd Wright and Buckminster Fuller. Then Magnus was invited to attend a mountain school in North Carolina. He became involved with an avant-garde young people's art gallery.

Magnus was one of the first black artists to open an art gallery, and he did so with his grandfather's money. Magnus's grandfather had made a fortune smuggling alcohol between the US and Canada.

Magnus was never without confidence. "I had a good childhood. When I was six, I learned to paint."

FROM A PROMINENT FAMILY

At the age of seven, Magnus began to draw and at the age of ten he could carve stone. He made a mausoleum for one of his Canadian uncle's wealthy customers.

This customer needed a family mausoleum within 16 weeks. Magnus's uncle did not have the time, so he told his client, "My nephew can do it for you." The client was reluctant. He looked at the ten-year-old boy doubtfully. This opportunity was a big step in Magnus's life. His uncle said, "I believe in him, I have faith, my nephew can do it." Magnus's uncle promised to do the mausoleum for free if his nephew couldn't complete the task as requested.

Magnus was in school and had to learn a great deal. Sixteen weeks were not enough for him to complete this enormous task, so he got a little more time from his uncle, but Magnus still doubted his ability to accomplish this project. "This is the biggest thing I have done in my life!" he exclaimed. His uncle remained unmoved. "Yes, you can do it, do it!"

Magnus designed and carved angels in marble from Vermont. The more he worked the better his work became. After 19½ weeks the mausoleum was ready. The young boy was excited and very nervous. The client was a very critical man. Magnus's uncle praised the boy's work and reassured him. With a bottle of Scotch, Magnus baptized the mausoleum. They covered the mausoleum with cloth before the customer came.

The customer arrived. "Are you ready?" the uncle asked. Then he pulled off the cloth. There were Michelangelo, Da Vinci – all Magnus's heroes were there on the mausoleum. It was a crowning achievement.

The man was so astonished, he looked at the young artist and put his arms around his neck. "If ever you need anything, here is my card. Phone me!"

Years later, when Magnus Harrell opened his first gallery in Buffalo, a man in a wheelchair came to the opening. For US $30,000 he bought the Black Madonna, the portrait of Olivia, Magnus's fifth wife, painted by Magnus. It was the man for whom Magnus had made the mausoleum. He was 102 years old.

Dr. Magnus learned a lot in addition to school. His mother's brother's wife taught him singing. He sang and played the piano. Magnus developed great self-confidence, even from childhood, because his parents and family reassured him. "My life was wonderful because I was loved." He had educated parents and many mentors. "Today there is such a lack of mentors around."

His parents had different people for dinner every night. Many of them were faculty from the university. "In a sense I have never been a child. I was always with adults. They talked to me like an adult. I never had a teenage girlfriend."

SCHOOL

Magnus went to high school in Buffalo and was Valedictorian. He specialized in Liberal Arts, and also learned Latin and German. Magnus skipped some grades at school. "I was far more advanced."

Magnus studied Fine Arts and Art History, Restoration and

Anthropology at Cornell University with a scholarship. "I was young, I was learning. I know who I am. I come from a dual concept." (Indian and black American) Companies were always looking for gifted, special people. "I was one of those."

MAGNUS'S PARENTS

Magnus's father, Cornelius Harrell, was an accountant. He and Magnus's mother met in accounting courses at the University of Chicago. It was a shotgun wedding. Later they moved from Chicago to Canada. His mother was 17 when Magnus was born. Magnus's father was an Indian born on a reservation. His father died at the age of 27, when Magnus was seven.

Ernest J. Lilly courted Magnus's mother for three years before she agreed to marry him. His grandmother arranged the marriage. Ernest had a lot of money and was a contractor. His dream was to become an architect.

Magnus was 11 years old when Ernest one day said to him: "I want to marry your mother; I will be your father." Magnus became rude. "I didn't want him to take my place."

There was great love between Magnus's mother and Ernest, his stepfather.

Years later, in 1962, Magnus's third wife died of appendicitis. Magnus was shocked. Every night he visited her grave. Then one night he saw a black spider. "It is time to go home," he thought. He traveled back to his parents' home and learned that his stepfather had died.

When Magnus was traveling in the US or traveling the world his mother often asked him when he phoned her: "When will you come home?" She wanted him to stay. "I can't, I have my business," Magnus always replied.

She died a few years ago.

PRISONER OF WAR IN KOREA

The Korean War started on June 1, 1951. "My job was to fly planes from Berkeley, California near San Francisco to Kobe in Japan. It was a shuttle service."

The United States Air Force needed extra pilots to drop bombs over Korea. They knew that Magnus could fly low. The flight pay was US $250 per day. Magnus did not care about the pay. "I wanted to show that I could." He flew for some time. Then he was shot down.

He was a prisoner of war (POW) for 3 years and endured a great deal of hardship. Recovery from the cruelties suffered during imprisonment took 4 or 5 years. Of several thousand prisoners, only 600 survived. Magnus weighed only 78 pounds when he was saved. "One must have something inside to survive."

FIRST GALLERY IN BUFFALO

When Magnus had recovered, he opened his first gallery. Then he decided to teach and opened his own school, "The Masten Community Workshop," with the help of the New York State Council of Arts. His school attracted students from ages 9 to 90.

His first gallery was small. He was working day and night painting, sculpting and preparing for exhibition.

The house had two floors. The sculptures he put downstairs. All the sculptures were draped, and also all the paintings.

On the night before the grand opening of the gallery, Magnus called the art critic for the Buffalo Courier Express. "My name is Cornelius Magnus Harrell. Tomorrow I will open my gallery." He invited the journalist to come that evening if he wished to have a cover story. "You have to take photographs. Otherwise, don't come."

The reporters arrived. "Take the drapes off!" "Of course." As soon as the reporter saw the pictures and sculptures, he ran to the telephone to stop the printing press. He had a new cover story. The next day the picture of the Black Madonna was in all the newspapers, to the annoyance of her uncle. Olivia was still living at her parents' home at the time.

That day, Magnus earned US $80,000. This opening was a sensation throughout the city. Magnus was so tired that a friend took him to her home and put him to bed. He slept for three days.

SECOND GALLERY IN BUFFALO

Magnus's gallery was soon too small for him. He moved into a huge three-story building his parents had left him. It housed a few shops as well as a school and a gallery.

Magnus moved to the top floor. On the second floor were sculpture exhibitions. There were also a dance floor and a classroom. The gallery was on the first floor and his school was in the basement. Magnus dealt with virtually all the artists in the city.

Magnus had a busy workday. In the morning he did 20-30 drawings spontaneously and quickly. From 3 am to 1 pm he did his creative work – his drawings and his paintings.

"Consistency counts in working spontaneously and quickly. If you work for yourself you have to do two days for one. I worked at least 12 hours a day, usually 16 to 18, 7 days a week. I don't have days off, I don't need them and I don't want them. American people today have become very lazy."

Magnus was often heard on the radio and seen on TV. With success came many friends and many enemies.

PICASSO AND OTHER FAMOUS PEOPLE

"I like my good friend Picasso. I met him and I went to his studio in Paris."

After Magnus opened his second gallery in Buffalo, he visited a friend in Paris. This friend asked if he would like to come with him to see Picasso in his studio. Magnus joined him and was introduced to the great master. Picasso asked Magnus, "Would you like to come back tomorrow?"

Magnus went there the next day. They drank a few bottles and Magnus learned several of the techniques Picasso used. Magnus learned a great deal from Picasso in those two days. One of the most important things Picasso taught him was to go forward.

Picasso used ten easels for his ten paintings a day. Magnus learned this from Picasso. Picasso worked as fast as he could. "Do as much as you can in a day, because tomorrow is not promised. Whatever you do, put your soul into it."

ATLANTA, GEORGIA

Magnus's doctor from the military, who had helped him so much after Korea, came to see him. He had come every year to buy a picture from the painter. When the doctor arrived, Magnus's wife, Olivia, told him about Magnus's situation. Magnus desperately needed a rest. They put him into a car and took him to one of the doctor's houses, near Atlanta. The house had not been used for a long time.

This house was on a main road. Traffic passed the house every morning from 7 am to 9 am, and then again in the evening. Magnus hung his paintings out and people came to buy. This is how Magnus became the owner of a gallery outside Atlanta.

A man by the name of Ray Robinson, who had helped Magnus to administer the gallery in Buffalo, took over the gallery and school in Buffalo. Magnus put all of his things into a truck and moved his belongings to Atlanta. The doctor stayed there with Magnus for three or four weeks. The doctor died in 2005.

A politician brought Magnus food from his restaurant and they became friends. Magnus stayed in Atlanta for two years. The politician was furious when he left. Although he hated black people in general, he liked Magnus. "You remind me of Sammy Davis Jr.," said the politician.

MARRIAGE

Magnus was married five times. He's always liked gray-haired ladies. His first wife was Greta Sternheim, a woman of German origin who was several years older than he was. The marriage brought social acceptance. She was a white American. The marriage ended in divorce.

Magnus then married Rita, who was also several years older than he was. They were married for 3½ years, then she died.

His third wife Ella was Texan and a few years older than Magnus. They were married for several years. She was the mother of his daughter, who died in 1989.

His fourth wife Josephine was just a little older, four or five years. She was the mother of his son. They divorced after a very short marriage.

Magnus married Olivia, his fifth wife, in the late 1960s. She was a photographer. He painted her, the "Black Madonna," and he sold this picture for US $30,000 at his first gallery.

When they married, he was more than forty and she was twenty-one years of age.

"With Olivia there was always this special something. In a partnership you should walk into the kitchen, take her in your arms and say, 'I love you so much.' A relationship should be passionate."

When asked which of his five wives he had loved most he said, "I loved most the Black Madonna. She understood me best. I took her to dinner and dancing in the country club."

Talking about his wives Magnus said: "They gave me the freedom to be me and what I gave was enough for them. I received so much and gave so little."

PHD IN MEXICO

Magnus moved from Atlanta to California, where he worked in the movie industry, restoring buildings for movie sets.

Magnus kept going south. In Mexico, Magnus attended school to study artifacts for three years. He lived in the mountains and he painted on the beach. He had five horses and took tourists for rides. He also earned a Master's Degree in Psychology and a PhD in Anthropology.

WHY HE FIRST CAME TO BELIZE

Magnus first came to Belize to collect a debt in the early '70s. A man from Belize crashed into Magnus's car in the US, and then invited Magnus to Belize to pick up a car to make up for the crash.

Three years later, after a detour through Mexico to earn his PhD, Magnus headed for Belize. He first stayed with a taxi driver. The taxi driver invited him to spend the Christmas holidays with him and his family at his huge family compound in the Lamanai area, a small village not far from the precious famous Mayan ruins. Magnus bought a horse and rode towards Lamanai.

After Christmas, Magnus went to visit the man he'd come to Belize to see. The man bought Magnus another car and they became friends.

When Magnus first came to Belize there were fewer people on the cruise ships. He played canasta and bridge with the women on these ships, and he used to win. In this way he met a great many people.

Magnus bought an old medical clinic and turned it into a studio, but the place was too noisy for him. He sold his studio with a profit and bought a place in Maskall, which isn't far from Belize City. Then, Magnus returned to Buffalo because his mother was ill, and to get his wife, Olivia.

Back in Maskall, he painted a great deal. He enjoyed diving and found many artifacts, such as old bottles. He brought them back from the lagoon and used them in his art.

Magnus experienced a strange life in Belize, but he knew that something good would come from it. He now has more friends here in Belize than in America.

After some time he opened another gallery, this time in Belize City. The Attorney General bought a painting and gave him a building, a studio in the city. Shortly after that Magnus met Emory King, author of a number of books about Belizean history.

Emory King took him out for lunch. "Why don't we open up a gallery in my coffee house?" Emory gave Magnus money for the furniture and painting, and Magnus designed and created the interior. They called it "Admiral Burnaby's Coffee House." Since then Emory King has become his patron, and they have also become friends.

"Belize is not a poor country," Magnus reminds us. "There are only poor people."

TRAVELING THE WORLD

Magnus traveled and painted. He went to other parts of the Caribbean: Jamaica, Barbados and Antigua. The new colors and shapes, the scenery, and the life in other countries all influence his work. He traveled through the Caribbean to experience the diverse cultures that he has captured in his paintings. He concentrated not on being a tourist, but on knowing the people.

Magnus was once commissioned by a zoo to bring the biggest anaconda he could find to the US, alive. His expedition took him to Monte Grasso, Brazil, where he found a giant Piguary. He brought it alive to Naples, Florida.

In Belize, Magnus met people from various parts of the world. Very often he was invited to come to see them in their home countries. He went to Africa, Kenya, Nigeria and South Africa. On several occasions he visited South America. Once he traveled to Hawaii where he had a girlfriend, a cartoonist, who he liked a great deal.

Magnus also traveled to Saudi Arabia. He went to the desert. Once he was offered rice with lamb's eyes. He ate a couple of these. "I just wouldn't do it again." The son of a desert sheik had invited Magnus to travel from California to Saudi Arabia. He gave him a beautiful Arabian horse. Magnus remained there for six months.

MAGNUS'S WORK

Magnus still creates up to 30 drawings or studies each morning. Magnus has studied the work of many famous artists, and their work influences him. Magnus's work displays an enormous variety of styles. The masters who influenced Magnus most are Kandinski, Rembrandt, Picasso, and Maurice Utrillo. His work includes Impressionism, Realism, Dualism, and Dada, and he does portraits.

For his paintings he uses pastels and a pallet knife. Magnus paints a great deal with a knife instead of a brush. This is one of Rembrandt's techniques. For his miniatures, Magnus uses numerous types of paper, cloth, wood, and papyrus – anything that will accept a mark. He has made professional drawings with colored markers, pencil, chalk, pastels, pen, ink, and lipstick. "Nothing is worthless; everything can produce a work of art."

He once made a drawing in a restaurant in Oxaca, Mexico with eyebrow pencil and lipstick on a white tablecloth. He asked two ladies for a lipstick and an eyebrow pencil, which they gave him. "That is life; you take what you can get."

One of the musicians in the restaurant owned a gallery. The next morning, this picture Magnus had produced stood framed in the gallery next door to the restaurant.

"I try to make my memory photographic. I am the camera. A computer ties me up, I don't like it." For his studies he uses permanent markers, fine and ultra fine, and watercolor pens. "I can do the same thing with cheap products."

Magnus carves wood. All his pieces are individualized in structure. "I am an industrial designer." He has designed furniture, such as a chaise lounge. He makes his modern furniture using small models.

Magnus has taught how to make tapestries and carpets out of remnants. The kids he taught gathered the pieces and made carpets. They understand what recycling is. "It is spontaneous work I do." He loves to make murals and to teach young people how to make them.

For his recycled art, he investigates any and all materials. "What keeps me doing this? It is my scientific research background. People are trained as consumers. I would like to go back in time."

He began his sketches and drawings in 1978 and is still

making them. Magnus is unorthodox in his works. "People don't understand what I am doing." He claims, "Everything that came out of me came by itself. I am extremely emotional."

During the most agonizing times of his life, he needed solutions. "I was too much concerned about the result of my work." As for his work, "I don't know how I did it, but I did do it." It is not possible to understand.

Magnus has also frequently worked with stone.

Magnus once had an afternoon show on television called "Mr. Painter's Hour." He taught the audience how to paint and draw at home. "You can play with color." He had children guests for his show.

Magnus often puts the falcon symbol next to his signature or onto his studies and paintings because his grandfather once asked him, "Why don't you use the sign they gave you when you were initiated into the tribe?" The Mohawk tribe, by the Canadian border, had given him the falcon.

"My creativity is a flowing thing. I know who I am and I don't need the title 'Master.'" His new style in 2006 was Minimalism, both in form and in color.

Magnus also likes to make pencil drawings. "Black and white, monochromatic makes people imagine colors. That shows you that you are doing a good job." Magnus likes to use acrylics, oils and pastels in his paintings. He loves his pictures. "These are my children."

BUILDINGS, HOUSES

One important new invention Magnus has made is houses for refugees that are folded into packages which can be dropped with parachutes into areas made inaccessible by catastrophes such as floods or hurricanes. The building is 12 feet high and designed on a frame concept. He uses plywood that can withstand severe hot or cold weather. It is cheap to ship, and an average person can put it together. Magnus has submitted a prototype of this house to the United Nations for evaluation.

Another of Magnus's great inventions shows his care for the poor and is designed for low income Belizean people. It is a comfortable house that can be built for only a few thousand dollars. "Why should someone have to pay so much money to have a house of his own?"

Magnus believes that you should not pay a mortgage all your life just to have a place to live. So he designs a variety of houses that cost between Blz $3,000 and 30,000 (US $1,500-15,000).

One of these great designs is the Tee Pee House. It costs only Blz $3,500 and can be put up in one week. It is made from recycled materials using a modernized Mayan concept. Standard panels and screens can be used. He made the first one for a writer. Altogether Magnus has built ten of these houses. Installing water and electricity costs Blz $6,000 (US $3,000), a very reasonable rate.

His architectural abilities were enhanced by his father and

his uncles in Canada, who helped him to learn art and sculpting when he was only a child. "I want to change the attitude people have towards architects, designers." Magnus has also designed and built cabanas, and he designed the Museum of Belizean History for Emory King.

MAKING SOMETHING OUT OF NOTHING – RECYCLING ART

Magnus tries to make people understand how to utilize indigenous materials, a knowledge the Mayan Indians had and the Mexicans have, but which was unknown to the Belizeans.

Magnus has designed furniture using lumber and sawdust or cardboard and sawdust, and shown others how to build it as well. "What one needs is to take a group of young people, train them in design, and show them how to produce inexpensive furniture made exclusively of indigenous materials like dead tree limbs for door arches and beautiful sculptures." Small stools for children can be built. Magnus makes use of everything he can find. Some people call him the "junk man."

On his property in Maskall, one day Magnus sank his hands deep into leaves on the ground. This material was good for raising plants. Magnus sold the leaves as fertilizer for Blz $100 per bag.

Magnus gave a series of lectures at Buffalo University on indigenous materials, recycling and conservation.

COMMUNITY WORK

Unskilled young Hispanics can be a problem in the US, especially those with a background of war, such as child soldiers. Magnus will approach a community, either in the US or Belize, and ask them for a deserted old building. This building is then renovated by young people, as Magnus teaches them the necessary skills. Then they paint murals on or around it. With this the community gets a better building, the surroundings are embellished and the young people can start to earn money for the art they produce. A demand for their work is created. Magnus did a great deal of this community work in the '70s, '80s and '90s. It was and is part of his career.

Just after 9/11 in 2001, Magnus volunteered to join a community project in Florida. The children were special children; some had fought as child soldiers in the Nicaraguan army and some of them had killed people.

Magnus received a studio, an apartment and a stipend. He helped unskilled mothers acquire the skills to open a small business. They sewed and Magnus showed them how to do tapestry. The buildings were repaired and repainted, and these high school dropouts became part of this new business: a design shop.

Now they generate income, the skilled teach the unskilled, and some of the older boys and girls work in administration. The business had a profit of US $19,000 in 2005.

These young people are selling their skills all over the country. They paint street signs and they decorate. They sell artwork to private homes, clubs and restaurants.

TEACHING

One of Magnus's great passions is educating young people. Very early in life he decided to pass on his skills and his knowledge to other people.

In Belize he has dealt with young people with behavioral problems. He has taught them industrial design and art, helping them to improve their skills and their lives. His studio in Belize is located in one of the poorest areas in town. From there he works with youngsters and young adults.

Belize City, the former capital, houses – among other things – poverty, unemployment and crime. Often there is explosive anger and aggression. The government supports Magnus's community projects.

One of the main problems in Belize is a lack of education. There are many criminals because of this. Magnus selected a group of young criminals and brought them into a youth hostel. He taught them industrial design. He also teaches young people how to make pictures out of sand and gravel and how to make frames. They can now earn money and also learn discipline.

Magnus teaches boys and girls. He teaches girls how to use tools. He teaches young women skills so that they will not be abused at home, such as being alert, concerned and observant. This way much aggression, which turns into abuse, can be intercepted and reduced.

"My classroom teaching is spontaneous, not classical, not traditional, but it works. The techniques work at different levels."

BELIZE

Magnus has invested much in Belize with his paintings, charity work and teaching. Nevertheless, he does not have residency, and he does not want it. He likes to come and go as he wishes.

The country gives something back to him. He feels he has something still to accomplish in Belize; and yet, he doesn't know what it is.

"Although the government says that the literacy rate is high in Belize, people are not always articulate and they are only semi-literate. They have difficulty filling out forms. The people tend to be superstitious. It is most interesting to meet people on a personal level. I am in touch with many people in the country.

"Women here in Belize want lighter-skinned children. There is no real independence here. The ruling party is in control. I am far better educated than they are. I can do things here in Belize that I cannot do in the US.

"I know the country, the jungle, the rivers. In Belize City there is robbery. Bad guys are there."

Prime Minister Musa asked Magnus to teach his son. Magnus taught him modern art. He is now the President of the Image Factory, the art gallery in Belize City. Some of Magnus's Belizean pictures have been exhibited there. Before Magnus, there was no modern art in Belize. "I like to make history. I understand how important it is making history."

GRIEVING FOR OLIVIA

Magnus gave his beloved wife the best care possible when she had heart trouble, but it was in vain. Olivia died at the age of 54 in Florida after a heart operation. They were married for 32 years. Her pictures are at his house in Buffalo.

After Olivia's death he was deeply depressed for a long time. He buried himself in work. He worked 16 to 18 hours a day with 4 hours' sleep.

During his grieving, he painted 10,000 paintings. This work saved him from sleeping. Sleeping meant dreaming horrible dreams.

Magnus wants to be more loveable now. The time he spent with his family was more important than he knew.

Also, too much administration in his life affected his creativity. "My life was always in chaos." He was hiding from people, but Magnus now admits that he needs people. "I don't like crowds, I don't like many people. I am a singular person." His wives kept many people away from him to make it easier for him to concentrate on his creative work.

"I am busy. I am at work." Sometimes he feels comfortable, sometimes he feels lonely. "I need a home, I need that more than ever now."

Ave Maria University in Florida offered him a position as a lecturing resident in fine art and anthropology in July 2004. But in August 2004 Magnus left again for Belize. He wanted to work with the kids in the youth hostel again.

CHARITY WORK

Caring for people is one of Magnus's main concerns. He is a member of the National Foundation for Transplants in the US. He has raised money for sick people who suffer from sickle cell anemia, a disease which afflicts people of African and Jewish descent. Another area of concern for him is heart and kidney diseases. He has raised money for transplants. He couldn't save Olivia from dying of heart disease, but he has helped other people to survive through transplants.

Magnus has worked for hospitals, senior citizens' homes and hospices. "I use color psychologically." Magnus emphasizes that he is of the opinion mental hospitals need to be more colorful, as do jails. "Mobiles capture the attention, movement is so important for the eyes, and so are lines."

POSTERITY AND MAGNUS: "I WANT THE NOBEL PRIZE"

After the years of mourning, Magnus is back to life. He gets up at three in the morning. He jumps out of bed and is ready to go. He starts to draw and make sketches.

Magnus wants to be remembered as someone who has done something.

Magnus has several book projects. He wants to write about techniques, theories and research. He wants to write a book about human relations. He wants to write about the many places he has traveled and experiences he has had. These next years are very important to him, from age 83 to age 90.

"I want to go against the grain. I don't expect rewards."

Magnus regrets that Bush took away all the liberties with the "homeland security" that Jefferson and Washington brought to the Americans. "You have to take care of freedom yourself. There is a social political apathy in the US."

Magnus believes he has been put here on earth to do something special. "All I do is for posterity and my fellow human beings. I have to do a job, I haven't done it yet." Magnus feels a call that brings him into the future.

"The Universe has always been very gracious to me since my childhood."

He wants to be remembered for his work. "This was my path from the time I was six years old."

MAGNUS'S PHILOSOPHY OF LIFE

"One of the great problems of our time is that people need to learn how to love one another. They often are married for decades, but they don't live together, they coexist." Magnus wants to write a book about partnership.

With his approach to life, community work, charitable work, teaching and producing art, Magnus has a certain philosophy. "When you give, there are people out there who make sure you survive."

Magnus believes that he has a purpose. He knows that by observing, by listening and hearing, and by experience, everyone can find the special causes in their lives.

"A strong Victorianism prevails. Constraints are still put on women. Often we are still back in our thinking some 200 years. I am trying to be warm, considerate and understanding. I love my freedom and my independence. My allegiance is to Magnus. I don't belong to any country, any culture."

Activity has a high importance for Magnus. "The doing is the important thing." He lives while working. "The real is in the doing." Magnus also likes to fly. He renewed his private pilot's license in Florida a couple of years ago.

Magnus has a strong sense of perseverance. He tells people, "Never give up." He believes in constant improvement. "When one advances we all advance." He has a strong vocation. "If I come into people's lives I want to make them better.

"I do a great deal of good for people because I feel I am blessed and I want to pass it on. When I do good something good comes to me.

"My mother always told me, 'No man is an island unto himself.' People like me. I always laid back and let the people choose me.

"I think I can walk if I try. Faith will move a mountain. People won't put a value on you. You have to put a value on yourself. You have to be what you want to be and not what people want you to be. I want to be natural, I want to be real."

Magnus has a strong sense of responsibility. "We have to do what we have to do." He strongly believes in vocation, strongly believes in his own vocation, and he believes that there is still another vocation for him here in Belize. He sees himself also from another angle. "I am an experiment in human relations." Human relations are an important topic for him. "I am a lone warrior."

Magnus sees the chances in life. "Opportunity knocks every minute, every hour, every day. People who do negative stuff don't see the opportunities. The universe demands that we are aware and look around and observe. We have to stop the garbage."

Magnus loves to encourage faith. "The universe will pay me." Magnus is convinced that "there is no coincidence in life." In his mind, belief creates a reciprocal effect.

ON OPRAH

It was sheer good luck that Magnus came into contact with Oprah. A friend of Magnus in Florida phoned Oprah Winfrey's company and asked her employees what kind of paintings Oprah likes. He professionally packed an impressionistic painting from Magnus and sent it highly insured to Oprah; she had to sign for it personally.

Oprah wrote a personal letter to Magnus to say that she is interested in his paintings. She wants to produce a documentary about black painters born before 1930 who had galleries in the US. There are three others still living.

FUTURE

Magnus is on his way to a new future. He wants to get settled in a well-built, comfortable home. There is a lady who would like to marry him. He likes her, but this is not what he wants.

Magnus can acquire scholarships in the United States to give them to Belizean students, allowing them to obtain a good education in the United States.

The years 2006/'07 have a whole new format for Magnus. He has developed a new style, Minimalism, and is working on an entirely new collection of work. He is looking forward to the Oprah project. Of course, he still wants to make studies, paintings and drawings.

Magnus wants to do an exhibition at the art festival in Cannes.

"A new period in my life, 2006 is a new era. I have survived and I am stronger than before. I like to share. I don't know what I will do this next year, but I have faith. There is a light on the horizon when one least expects it to be there."

Magnus wants to write a book: "Magnus the Magnificent."

HIS NEW HOUSE

In May 2006 Magnus rented a two-story house in Belize City to open a new school and art gallery, as well as for his residence. The town has entrusted him with a project to teach art and industrial design to young people aged 14 to 21. This will be a new cultural and social center in Belize City, and a place where young people can become better integrated into society.

If you are interested in Dr. Magnus's art, please contact (helga.peham@chello.at).

Magnus Signing a Study

Dr. Cornelius Magnus Harrell with a Painting

Harmonies...

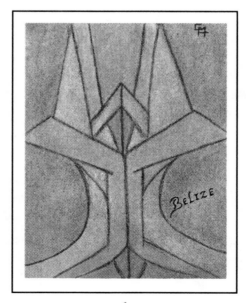

of...

Lovely...

Belize.

PART V: SCHOOL AND UNIVERSITY

PART V, CHAPTER 28: AESTHETIC PHOTOGRAPHY

Professor Crandall Hutchkins co-founded an art center in Concord, Massachusetts. He was able to rent the building for a few dollars a year. This was a major project. When he left there were over 100 artists there.

He was born in 1942. "People born in the '40s are lucky people. We had good times and great times, a good childhood."

He is a professor at the Massachusetts College of Arts in Boston. He's also an artist, and he has a degree in Art and in Philosophy. "I think of myself as an artist who studied philosophy."

In fact he has five degrees. He has a doctorate in Philosophy, specializing in Aesthetics, from one of the world's most prestigious institutions of higher learning – Harvard University. In addition he has Master's Degrees in Fine Arts, Photography, and English Literature, and has been teaching these subjects. First he taught at colleges, then at art centers. "I am big on education."

Professor Hutchkins taught at the Emerson Umbrella for the Arts, an art center in Concord. Many artists are there. Writers have studios, painters have studios, and many interesting people work there. He has also written grants to help people, especially the needy. He is a prominent man.

His teaching is varied. Professor Hutchkins taught a course

in the Tao of Clay. Tao is the world's most ancient Chinese philosophy. "I wanted to create a course to teach students something about the nature of making things. Tao and clay work really well together."

For many years, he had many students. "They had to stay with me for at least three years. I wanted to get away from traditional teaching systems. Art needs to be taught in a different way, to develop a community. I had a lot of interesting students; it was a wonderful time, it is a very great place." Crandall decided that he wanted to start photographic projects, and left five years ago.

His artwork is three-dimensional and two-dimensional. He has made clay sculptures, paintings, drawings and photography. Now he spends most of his time working at photography. "I have been doing a lot of photography these past years." Most importantly, he has documented the island of San Pedro in Belize in thousands of pictures, some of which appear throughout this book.

Professor Hutchkins and his wife came to Belize and San Pedro for the first time in the 1970s and '80s. She likes to snorkel but she doesn't like hot weather. "That is why she loves Paris in the spring and in the fall. I love coming here because of the colors and the people. My wife Beth loves to sit in a café in Paris and go to the opera."

Crandall first came to Belize in 1961 with his uncle. He went swimming a lot. His uncle used to have a place in Key West and they went to Cuba and came down to Belize. "I was a teenager; all was romantic, very exotic."

But he didn't return until around 1978. "We came a lot in the '80s and '90s. Then my wife said no."

Crandall came again to San Pedro in 2001. So much had changed. So many different people lived on the island. Before, there were only the Mexicans.

"Mrs. Gonzales – her grandson runs the medical clinic – is the oldest person on the island. She has seen the island develop from a few hundred people until now. Her mind is very sharp. She came here as a teenager, before a hurricane, and is an amazing person to interview. I was fascinated."

In 2001 he decided to photograph the island of San Pedro. He wanted to make a visual document of historical things that will not be here in another 25 years. That way, the grandchildren of people living here now will know what the island was like before. He made the photographs in order to preserve the beauty of the island for generations to come.

He will give the digital photographs, and his writing, to the Belize National Library. He started the project in 2001 and finished it in 2006. To spread the information, Crandall plans to also create a website with his photographs and writing. He now needs to combine the valuable material he has collected during these five years with the many mini-interviews he collected from people living on the island.

Next year he will start a new project, photographing Saudi Arabia, Palestine, Jordan, Lebanon, and Israel. This is a much more difficult project, one which he expects will take three years. Saudi Arabia is a difficult country to get in and around. "My wife does not like the idea. She would prefer that I photograph Paris. The Middle East frightens her."

He teaches at Rhode Island School for Art and Design. "I take classes there whenever I can. I go from film to digital."

Before 2001, the demographics were very different in Belize. Most of the island was populated by people from the old families of Ambergris Caye and by mainland Belizeans, people of the country. At that time everything was rather primitive. Electricity, water and the infrastructure were not developed as they are now. "There was virtually no crime here. You could leave your laptop and it would be there tomorrow." Land was very inexpensive and undeveloped ocean land was US $10,000 per acre.

Then the hurricanes came, bringing many people from other places. They stayed. Golf carts came. New buildings and also tall buildings came. They were very different from the old wooden structures. "Many people are here now; it is very crowded here compared to before."

Conch shells and lobsters are prevalent shellfish. Now there are many tourists coming to the island on a holiday. San Pedro depends on tourism. Everything is imported on this island. It is becoming more expensive, also for the people who live here. Medical care is better.

"I want to make a record of the things that are disappearing. Many of the things I photographed are already gone. The teapot in front of Ruby's Hotel is gone. I want to capture the things that won't be here in the future. If I come back to San Pedro I know many of the wooden buildings will be gone. I photograph things that I want people to see 50 years from now. I love this place. It is a very beautiful place. It is sad to see things go away, disappear. The new bridge opened recently and it is wonderful, but the hand-driven ferry is gone, and it was also wonderful."

San Pedro was easy to photograph because Crandall knows it well.

"You cannot just go to a place and photograph it. You have to learn to know and to respect a place before photographing it. Before I made snapshots, so when I came I knew. Then I had a schedule of work. I got up at 4 o'clock in the morning, before light comes up, so that I could capture different lights. Every day I went 15 miles a day. I worked until 11:30 or 12, then stopped for a couple of hours and worked until 7 or 8 at night and got home very tired. That was my schedule. It was a wonderful experience. While working I met so many great people."

Crandall approaches photography as a visual essay. He tries to understand the place and the rhythms, see the people, hear and rehear.

"I feel a lot of responsibility. Nobody photographed this place in a serious way. Therefore, I had to do the best job I could do."

Crandall wrote to the Government of Belize. He wants to keep the copyrights but will donate the pictures for educational purposes. He made 15,000 digital images in three years. "I want to keep control of it. It is a lot of work. I feel very privileged to do this. Two thousand pictures will be selected and edited. San Pedro on Ambergris Caye is a very beautiful place."

Selecting the images, and editing them to make them the best possible, will require hundreds of hours. He used an

Olympus Mamiya digital camera, medium format, large negatives. The images are all numbered. Crandall's comments on the images are on a small digital recorder. Everything is digital, which is very convenient, less polluting to the environment, and great fun.

"When I get home I need to clear my mind for some weeks, stop for a while. I have been working on this project at an intense level.

"My wife would love to go to Paris as an expatriate. I would always have to go back to my roots. I'm a diabetic. I love to swim, to exercise. We are thinking of moving to San Francisco or Vancouver. San Francisco is like a European city with a lot of cafés, a lot of cultural activity. Certainly a place that is not cold."

PART V, CHAPTER 29:
SUNSHINE AND MOONLIGHT

Professor Floyd Jackson, a psychiatrist, a Yale and Harvard graduate, came to Belize after many years of practice and teaching in the US, to accept a position as Professor of Medicine and Dean of Students at a medical university in San Pedro. He accepted the position over other domestic ones because he wanted to live somewhere exotic with a sense of adventure.

WELCOME TO BELIZE

"I went to the Caribbean several times, and to South America, but not to Belize. So I came here sight unseen and thought it would be an adventure and exotic. So it has been pretty nice. The sea is nice, the sandy beaches are lovely, and there are a lot of interesting people living and vacationing here."

A week after his arrival on the island, Hurricane Keith, a Category Five, struck and wiped out the medical school. "We had been told by the US Embassy to evacuate but Dr. Jeff, the president of the medical school, made a judgement call to the contrary. The hurricane came around noon, passed us, then turned and hit from the other side, and arrived at about midnight with full force."

From his condominium Floyd witnessed an 18-foot surge from the sea that completely submerged the swimming pool at the residence. He was sealed in his condo for three days due to the force of the storm.

"There was no food, no water the first few days. Then the army from Great Britain came and some of the Canadian Army members came with food. Great Britain came first with food. The US did not help. I was really upset with that. No water, no electricity for three weeks. I was fortunate, because Royal Palm was solidly built. Spin Drift Hotel became the clinic. We all went there to help. Eight hundred people needed suturing. Flying debris had hit them. There were cuts to be sewed. Most people on the island lived in places that were destroyed by the hurricane. There were some dead bodies on the boats. Nobody died on the island, except on the boats."

After that, the school was evacuated and Floyd was sent to Orlando, Florida. The University of Central Florida was there, and St. Matthew's University (his school) was able to use a building or two.

"I taught accelerated schedules because we had lost three weeks. Then I went to Maine to teach all the other students, second year. In Maine I also had to teach classes that I wasn't hired to teach."

Several of the teachers could not come. Floyd taught Pharmacology and a few other courses. They followed a very busy schedule, about 10 hours a day, because they had to have a certain number of hours for WHO accreditation. This was in September and October 2000. Dr. Renae, the dean, was in Orlando. Floyd helped her there with administrative matters as well. At that time they had about 500 students, and they also had to find places for all faculty members.

BACK TO BELIZE

After one year he returned to Belize, in September 2001, to teach for another year. He taught four-day weeks. He liked returning to Belize, although Maine was a pleasant rural area and he could have stayed and taught there. The hurricane didn't affect enrollment.

Dr. Jeff had new buildings, again pre-fabricated, constructed in San Pedro. These were all-wooden structures on Banyan Bay space. A crew assembled parts that had arrived from elsewhere, but the second time they built them a bit more solidly. When the school left the island in 2003, they gave the buildings to people in San Pedro and stored the contents at Royal Palm.

"When I came here I had to look for a house. Many properties were for sale because they were damaged, and you could get them very cheap. All professors were getting well paid, six-digit incomes. So I bought a house in the San Pablo area of the town.

"After the hurricane, people showed that they are of great backbone because a lot of women and men lost everything. They were cleaning up rubble, were not just depressed, but actively trying to start all over again. So I found that the people had a certain resiliency. You saw women cleaning up rubbish, trying to save something, men making new structures. They tend to be a happy people. They are very very poor people. They have a sense of community, of strength.

"The people tend not to be very educated or very cultural compared to first world standards. They tend to be rich virtually, but education wise there is a deficiency. They tend not be ambitious. But of course there are exceptions. Most are very family-oriented. If they have enough to live on they don't really want to travel, just want to be workers, have a family, live on the island. People like it here; it is a peaceful place. Lots of families have been living here for a long time, a lot are related, some common parentage, and great grandparents – interesting people.

"It is hard for people to make money here. They have to do tourism or some business related to tourism. That is what people do. Some time ago they made a living by fishing, now they take people out fishing. Business tends to revolve around tourism. They tend not to want to be proprietors, but are happy to work for proprietors from the US, Canada, Great Britain. They are not ambitious. They need education, business knowledge. But, of course, there are exceptions.

"Also, Belizean people don't have many resources, not much money. Few Belizeans are proprietors; they are often into drugs, and have a drug business on the island, diving business and drug business. Most of my friends tend to be tourists."

Floyd doesn't find that the Belizeans have a stimulating background; he can only be superficial with them.

"We don't have common interests and common backgrounds. But there are numerous tourists. I meet interesting tourists and also expatriates. Some are very interesting. Most other expatriates are not so friendly with Belizeans. For example, at Fido's virtually all customers are expatriates or tourists. Expatriates and tourists socialize more with one another than with Belizeans. Belizeans socialize among themselves. Now I just met a couple from Iraq who have come here on vacation.

"The most interesting tourist I met was a prince from Austria, Raphael Orsini-Rosenberg. He was very fun. I was playing some Viennese waltzes on the beach. He was coming to me and saying I was playing music from his country. I said you must be from Austria. Yes, he said, he was from Vienna. We went to lunch together. He had checked into Ramon's Village, a lovely resort in town. He looked like a backpacker, not like a prince. He was very cultured, very educated. He has a degree in French literature. He told me a week later that he was a prince.

"I had just moved from my house on the yacht. He wanted to see it. He liked it right away and smiled. 'I could live here easily.' 'I have an extra bedroom on my boat,' so I told him, 'If you like, you can stay here.' 'I'd love to.'

"So we went to Ramon's and got his things. We went fishing and had good food to eat. He loved the bakeries. We went snorkeling. We went on a small boat to the reef, took a tour around island, paddled to some small islands and did bird watching on one island. We were traveling on a boat all around on day trips and we used to have nice conversations most evenings.

"When I mentioned that Prince William had been here, Prince Raphael mentioned, 'I know Prince Will.' They were friends. I wondered how that could be. Then he told me his background. He was a fascinating fellow."

Floyd has a very open personality. He easily talks to tourists when strolling along the beach, going through the streets of

San Pedro or sitting in the CocoNet Internet Café, a good meeting point in town. Everybody seems to know him. You often hear "Hey Doc!" when you walk with him downtown.

"I met remarkable people from Germany, France, Spain, Italy, several from Mongolia, Nepal, Siberia, Russia, Ukraine and China and Japan. I met a very noteworthy fellow from Japan I am still in contact with. What's so unique about him and our friendship is that he couldn't speak English and I could only speak a few words of Japanese. A lot of sign language and teaching each other helped. We were just inseparable for this week he was here; he lived in one of the hotels. We met each other every day, discussed things with sign language and taught each other English and Japanese. He was a young fellow. His name was Aki-Yori. He was traveling alone. I met him on the beach. You meet so many interesting people on the beach."

When the St. Matthew's School of Medicine relocated to Grand Cayman in 2003 Floyd decided to stay on the island.

MORE STORIES TO TELL

"I met soldiers from Singapore. They looked quite exotic dressed in their uniforms. I showed them around town. It was a relationship for just a few hours. They spoke English very well and wanted to see something different. They were doing jungle survival training in Belize."

Floyd has met a lot of British people as well. Belize belongs to the British Commonwealth.

"One man was a helicopter pilot. He came from England. A lot of army people come here to do their training in the jungle. The British has still army bases here in Belize, on the island St. George's Caye, near Belize City.

"What keeps me here is I meet so many interesting tourists. And also, I socialize with a lot of fascinating expatriates from all over the world who decide to live here, such as Miss Kate, who helped out with the Music School, a very good teacher from Canada.

"Steve is from Connecticut. He gives me Internet time. Walter, who is from the Ukraine, also lives in America now. My friends are essentially expatriates.

"I still have a dream of having a performing arts center, but need to find financing for it. It would be a great center. It would serve a wonderful purpose by featuring Belizeans, and also for expatriates and tourists this center would be good. I think it would be a wonderful mechanism to expose Belizeans to high culture and education. There is so much education in dance, ballet, opera, acting, film, music, classical music, symphony, and chamber music. It would be so enriching for San Pedro if they had this performing arts center. I meet the artists in various ways. They give me enrichment in the arts. It is just a need that I see. A lot of people come here, but they would never choose to live here because of the lack of culture."

PSYCHIATRY

Floyd is one of three psychiatrists in Belize. The other two are Roy Lopez and Dr. Claudina Cayetano, at Goldson Hospital, which is a mental hospital. Both are black, both are Belizean, and both are Garifunas. They both go to Garifuna ceremonies in Dangriga, do the Garifuna dances, and work very closely with shamans.

Floyd learned transethnic psychiatry at Harvard.

"The Mayans worked with natural herbs, but they also pray to the gods. So culture is not separate from the psychiatry of working with Belizean patients. You have to appreciate the culture. You have to have cultural sensitivity."

CREATION

"In addition to psychiatry I do my own art projects. Composing, I used to perform for the church, I did that for three years. I used to do music recitals at the Brown Sugar Theater Cinema, which is no longer there. People here don't go to the movies. I write poetry. This is a relatively new project. Music composition I have been doing since I came here. I studied composition for five years; my degree is in composition and violin. Here I have the freedom to do composition. No strain, no stress, I feel free to compose."

Floyd started the Free Music School of Belize and received a lot of support from the Governor General of Belize, Cobell Young, who champions any music performance in Belize. His son is Cobell Young Jr.

"We have coordinated music efforts for San Pedro. [Cobell] Young [Jr.] is the music teacher at Palotti High School in Belize. He has the only orchestra in Belize, an amateur orchestra of high school students and some adults who perform with them." Floyd loves to perform classical music.

MANY INTERESTS

Floyd lives on his yacht and walks his dog. He also likes diving and snorkeling. Belize is a diving Mecca of the world; the Barrier Reef is the largest reef in the Western hemisphere, the second largest in the world after the Great Barrier Reef in Australia.

"I still go diving. There are different levels of diving. I like to take courses. With each course you get additional dives. I've earned three additional licenses here in San Pedro. Fourth level would be instructor. I am a dive master here."

Floyd wrote a column in the newspaper for a year. "Like Dear Abby. People would write questions. 'I am fighting with my husband, what shall I do?' 'My son is very depressed, how can he get out of the depression?'" Floyd answered them weekly and explained mental health, which is a very foreign idea in Belize.

Floyd volunteered at San Pedro High School for a year. One of the local kids hung himself. The parents were heartbroken. Now Floyd is on call. Locals he treats gratis.

Floyd also used to work for the Island Academy. Barry Bowen's wife, Miss Dixie, is the owner of the school. He still gets invited to their annual Christmas Party.

"The Island Academy compares to if not exceeds the private schools in the US or even Europe, England. It compares to Elite Schools.

"The people are very peaceful here, very little violence in public. There is domestic violence, a fair amount of that, not just among the locals, but also among the expatriates."

A WEALTH OF OPPORTUNITIES

"I think your best job here is if you are your own proprietor, if you could link into a computer type business, even hardware, jobs over the Internet, own a computer industry, perhaps supply people with hardware, software, and laptops. You cannot get any software here. My interest is in publishing; you cannot get the software here. You can't get MP3 software here; you can't go to a local software shop to get the software. A software and hardware supplier is needed. You can't even get business software here in San Pedro.

"Also computer education going into the schools would be important. A computer school would do very well. Teach how to use certain software packages, supplying software and teaching at school. It would be good to get a computer college here, computer sales, computer business, computer education, and working over the Internet. But this is third world. It would be important to have stipends, scholarships for locals. There are a lot of bright kids."

Floyd thinks tourism is focal. It is the ultimate industry in

Belize.

"Virtually every job on the island is tourism related. It is limited what you can do from outside. You need a tourist license. You can employ a Belizean with a license, work cooperatively with a Belizean. The Belize Tourist Board gives the necessary information.

"Live upstairs in a house and have a restaurant downstairs. You need to have some kind of specialty. Restaurants are hard to do; there are a lot of restaurants here. If you have a special idea for a restaurant, this would be good.

"One thing that is missing is sailing. There is plenty of opportunity for day sailors. There is not one sailing school. Also somewhere tourists can rent sailboats by the day, a few hours at a time.

"One other job would be a doggy hotel. One woman has a dog shop and also walks dogs. People might want to leave for a while, so a hotel for dogs would be good.

"Start an air school. John Greif owns Tropic Air. Someone could come in as an instructor. They could start an air school. You can get planes relatively cheap from Ebay. Also training for mechanics from Europe and the US. Set up an aviation school. Some people may want to fly, but have no place to rent a plane. John Greif has unused planes, but doesn't rent them.

"Education is the key.

"We need a health club that would provide yoga, dance, aerobics, cycling, jogging, cycling clubs, family bicycling trips.

"There are not enough boat mechanics. There is no

mechanic school for boat mechanics, or for golf cart mechanics. Nowadays even auto mechanics are needed, now that there are not only golf carts but also more and more cars here in San Pedro.

"Someone should set up a trade school where they have specialties: auto mechanics, boat mechanics, graphic design, things that are practical. High school kids finish school and have nowhere to go. We even need a language school. So many hotels cater to tourists, often Europeans, but elderly people don't speak English. Tourism is so big."

Floyd would also like to see someone in Belize with media experience, in radio and in TV.

"It is a very attractive place for business. You pay virtually no tax here, some 6% business tax."

He thinks that contractors are needed to keep up with the demand for new buildings. Standards are very lax when compared to the States.

"What they also could use here in San Pedro is a good photographer, for graduations. There are a lot of sites about Belize; people read the Internet."

FINAL THOUGHTS

"At the US medical school which was located in San Pedro until 2003, the students couldn't fail. That was a strange standard. Here it was a game; therefore many people from the faculty left because of these things. They couldn't flunk them, so they lost the students. I had brain damaged and senile students, really a strange game. There were a few good students. They were all from the States, even the Indian or Asian or African students. But it's nice to live here. It's very pretty here, the weather is sunny most of the time, and it is relatively safe. I like it.

"There are different styles of living and different amounts of money you need. If you budget it can be very inexpensive. There are a fair amount of people here who are alcoholic or drug dependent or running away from something. More and more people are coming. Many expatriates come. One came and bought an art frame shop. Also, the island is built. You could eat on US $10 a day, although maybe not your choice.

"There's no stress here. It is relaxing, a nice place to retire. Newspaper articles report that it is among the ten most desirable places in the world to retire. It is conducive to being creative here. I meet people from all over the world. People on vacation are always very pleasant. It is very harmonious, people get along. I have been able to make a lot of music here, hundreds and hundreds of poems. It is so stress free."

PART V, CHAPTER 30: CHILDREN AND CHOCOLATE

NOVA SCOTIA

Kate (Catherine Nash Crossan Whitney) is a teacher. She was born in Nova Scotia, Canada, in 1948, the year the United Nations was formed. She lived in a town of 3,000 people, in Annapolis Valley, an apple-growing district.

"I have an affinity for the sea, for water." Nova Scotia is almost an island, and Kate was attracted to Ambergris Caye in Belize.

TWO TEACHERS OPEN TO THE WORLD OUTSIDE

Kate has two Bachelor's degrees, one in Psychology and one in Education. Kate married the day after her 22nd birthday. Her husband Dave is also a teacher. The couple has two daughters, one born in 1973 and one in 1979.

In Nova Scotia, Kate and her husband subscribed to a deferred salary program. The Teachers' Union keeps 25% of the teacher's salary for four years; the fifth year is a year off with full pay. "That is how we reached Belize on a year off with full pay."

HEADING FOR BELIZE IN 1983

In 1978, their first deferred salary year, they left Montreal for a year to travel around Asia and Europe. When they returned to Canada, their first child was four years old. "Let's have a second child," Kate said, "And when that child is four years we will go again."

The family had a big atlas which they would look through all evening long, deliberating over the next place to visit. As a mother, Kate wanted a secure coastal place. It also had to be tropical and rural.

"I chose Belize. That is how we ended up here in Belize, in September 1983. It was a lovely trip. First we flew to see New York City and Miami, then we caught a Taca flight to Belize City. It was the rainy season."

When they reached Belize, the rain was pouring down.

"I experienced 'culture shock' for the first time and felt uncomfortable. We wanted a budget hotel. We are budget travelers. We went on foot with big backpacks, in the rain through the open sewers of Belize. I tried to keep my daughters' feet out of the contaminated water. I thought, 'What a big mistake.'"

They found a room in a budget hotel. The shower pipe was hanging out of the wall and there were dog feces in the compound. Dave said, "Don't worry, we'll go to one of the gringo hotels." They packed up and went to a good hotel and settled down for the evening.

"'Let's explore Ambergris Caye, the island, a new destination,' my husband suggested. I was still feeling anxious, but not verbalizing it. We went to Municipal Airport and arranged a flight there. I remember the pilot's name, Mandy, a delightful man. I was sitting beside the pilot, my 4-year-old daughter on my lap, my husband and second daughter seated behind us. I thought, 'I am an unfit parent.' The door didn't stay closed. The pilot reached continuously to secure it during the flight. My child was perfectly happy, cheerful, chatting away. I did not pass on my anxiety.

"I remember the moment of touchdown in San Pedro. There was a tiny office where we were greeted warmly by Iraida Gonzales and her son, a lively and affectionate 5-year-old child. It was her little travel agency, maybe the first on the island. We walked in with our rucksacks. My heart melted. I relaxed; I knew this is what it is. A face, a smile, a mother and a child. That is why we travel, to connect with other cultures and people.

"The lady said, 'Leave your stuff and your children here.' We explained that we would be here for nine months and as a family on a budget we needed accommodation, some place to stay and left it up to her. She was so knowledgeable. 'You will be comfortable with Philippe at Lilly's Hotel.'

"We were a novelty arriving with two children for nine months on the island. Right away we felt comfortable and accepted. I find as an entire culture, Belizeans have grace. Belizeans are so kind; they have charm, a warm smile, gentle manners, and restfulness. I love the Belizean people, whether Creole or Hispanic, they are a beautiful mix of people. I am delighted with the Belizean mentality. I find these people open, and generous with their time, lovely and relaxed in a leisurely way. They have that pace."

The family later moved to Sam's Hotel, where their room had a fridge and a stove.

TESS'S SCHOOL

Kate and Dave enrolled their elder daughter, Tess, who was 10, in San Pedro Roman Catholic School. First, though, they let her play on the beach for a week. At the time, the school had only ten classrooms and stood on stilts.

"The education system in Belize is based on the British system and is very demanding. Tess had to learn the names of every member of the government, the ministers, etc. There was great emphasis on language, grammar, sentence structure and vocabulary; it was a challenge for her. It was a team effort. Homework was intense. They have a school uniform that has to be cleaned and dried daily.

"The parents here want drills. They demand that their child has homework, in the Roman Catholic school. The curriculum is daunting. My thinking on it is that there is an imbalance. I feel the children should write more essays, should be reading far more. There is a lack of art and music at school, artistically and musically there is a vacuum. It is duller, more work driven than creativity driven."

Kate and her husband are Protestant, and they didn't know that Tess was expected to go to Mass every Sunday. So Tess had to sweep floors in the school. Kate watched her daughter sweep from a distance.

"All these kids were sweeping happily. The sand came back inside with the breeze. I was amused by their antics."

FRIENDS IN SAN PEDRO

"Ovidion Guerrero and his wife Elia operate Martha's Hotel, and now Martha's ice factory. They were our good friends." They met through their daughter: Anel and Tess were inseparable friends.

Four-year-old Jenna and Ilda Gonzalez were very close friends too. Dave and Kate befriended Ilda's mother, Iraida Gonzales, the first travel agent for Tropic Air. She arranged flights, accommodation, and travel for tourists.

"Iraida introduced us to her mother Miss Elvi of Elvi's Kitchen. I admire her tremendously. My children called her 'chi chi,' which means grandmother. Every day Elvi cooked for the fishermen and boys. They paid her a weekly amount. We arranged to have our noon meal there every day except Sunday when she was closed. On Sundays she accepted us as members of her family, too. Miss Elvi has four daughters and two sons. She was raising six children and running the restaurant. She fed a lot of people in construction, dive shops; locals and tourists. She made her own buns every day for the burgers. I love her.

"Her husband was a beautiful man, a very kind, affectionate man. He made hot chocolate. He was so loving with our children. He asked my daughter at the age of ten to take money to the bank teller. Tess would run barefoot to the bank with cash in a zippered bag to deposit.

"My husband's friend is Miss Elvi's son-in-law Dimas, a fine jeweler. Dimas had a license for black coral. He dove for black coral and would make jewelry from it, also from conch pearls. He makes original, artistic jewelry. Years of diving damaged his eyes, but in 1983 he was an active diver. My husband would chat with tourists and show them pieces for sale.

"Dimas was married to Gloria, Miss Elvi's second daughter. Their 4-year-old daughter, (little) Elvi, was a year older than my daughter. At that time Gloria was a primary school teacher. Now she manages the restaurant. Her sister Alma was a finalist in the Miss Belize competition. Another sister, Jenni, owns Caliente Restaurant. She followed in her mother's footsteps and opened a restaurant, specializing in seafood.

"Most gratifying was to be so comfortable, so effortlessly accepted into the community. Children are great ambassadors. They facilitate interaction. It just expands. It is a ripple effect."

FAMILY ROUTINE

"I found that the women were so busy. My husband could socialize all day, all evening. I couldn't socialize with women over a beer. Women are too busy to socialize. I read a lot, enjoyed solitude. I went to bars on weekends. I like to socialize with people. There were a lot of soldiers, British soldiers, and I enjoyed the conversations. I would go out on Friday, Saturday evening, engage in conversation, also with tourists. A place like this can also attract thieves, liars, interesting people, colorful characters, eccentrics, alcoholics.

"I fell into a very relaxed and happy routine. The children were comfortable, happy, and free. We didn't worry about them after we began to connect with people. They had to be home before dark, had to be home at mealtime. So in the evenings I really enjoyed reading to the children, doing homework with them, having family time. We joined the library here. We brought textbooks from Canada. My daughter could progress in her studies on her own."

EASY ASSIMILATION

The family of four soon felt comfortable and accepted. "The assimilation was very easy." Tess picked up Spanish quickly. In school all instruction is in English, but the children speak Spanish when they are socializing, associating, playing, or at homes with Hispanic families. "We were invited to share Christmas meals, birthdays. People were including us in family gatherings, being very generous to us. We also did a lot of swimming. I had met my husband swimming, when we both were teenage swimming instructors. We both love the water."

CHRISTMAS 1983

"At Christmas '83 I put a coffee tree branch with red berries in a coffee tin and stabilized it with a piece of coral. We had hand-painted wooden birds from Guatemala. We only had candy on the tree and these painted birds. We offered pretzels and candy from the tree to the guests."

During this nine-month period, they had to leave the country twice to get their visas renewed. For this reason, they took a holiday in Merida for a week, and later made another trip to Guatemala to have their passports stamped just after Christmas '83.

EASTER 1984 IN PLACENCIA

Easter '84, Kate took the bus to Placencia and rented a little room on the beach. There was no key, just a rope on a nail to secure the door. She stayed two nights and explored the area, looking for conch pearls – white, pink, yellow and orange. Her husband Dave would take them back to Canada and have them set in jewelry.

Dave took part in a manatee hunt, a kind of a tribal rite for the men in San Pedro.

"In the dead of night several men stole away and went to the lagoon side. Before dawn everybody went with pots, containers, plastic, when the manatee was butchered, carved up." Kate's husband brought a piece of manatee home, and Kate pressure-cooked it. "It tasted like pork. That was a highlight for my husband, that whole ritual. This has been happening here for centuries. He was grateful to be included."

Kate also ate turtle. "It tasted like beef. I really enjoyed it." Now Belizean laws protect both manatees and turtles.

BELIZE TRIPS (1984-1986 AND 1999-2004)

Kate and her family returned to San Pedro during school holidays in March 1985 and 1986 for about one month each time. Those two revisits were exhilarating for Kate, and all enjoyed the renewal of friendships. Each time they stayed at Sam's Hotel. Kate took home a watercolor painting, featuring a clothesline in a typical family yard. Dave chose a machete for his souvenir.

In 1999 Kate returned again to Belize, this time for a six-month interval, December to May, and since then she has returned annually, reaching Belize around November 1 and her home in Canada in early May.

"My husband accepts my lifestyle and my choices. We are both comfortable with our arrangement. I have a very strong character. I needed to get out to work, do my profession. I compromised, I only taught part time, as a substitute, so I don't qualify for a pension yet. I have never had continuous work. In 2008 I will reach 60 and early pension, which will facilitate my life here. My husband will buy me a ticket when I want to see my children in London.

"I am a Gemini and I enjoy two frames of reference. I have a high energy level. I am better diluted."

BACK IN SAN PEDRO 1999

In 1999 her youngest child was 20 and doing well at university. Kate was free to pursue her own interests.

"I left Canada in December 1999, 51 years of age. My intention was to stay six months. I brought 3,000 or $4,000 Canadian with me [US $2,500]. I was apprehensive, thinking, 'I haven't been to Belize for 13 years, it is going to be expensive.' I had a budget; it worried me. First I checked out Caye Caulker; it's cheaper. Two days later San Pedro was calling my name. I reached the dock, left my backpack with the police and walked to the beach in search of accommodation."

She went to Sam's place and saw Segundo, Sam's brother, working on his boat. He recognized her after 13 years. Kate spoke with Sam Junior.

"I wanted a room for six months and wanted to volunteer in school. I said, 'Give me the room for $100 a week and I can tutor your children.' Sam agreed. I was very relieved and thankful. I went to school with the boy, Brandon, two miles south, to the Seventh Day Adventist School New Horizon. I was tutoring and gave the teacher ideas. I passed a lot of strategies on to her and we formed a good partnership.

"I put the information out that I needed a bike, a cheap one, an old bike. The wife of one of the fishermen was pregnant. I paid $60 for her bicycle and agreed to give it back after six months. Five dollars per 7-10 days, a very strict budget."

Kate made soup from fish heads and bones discarded by the fishermen at the pier, and thickened it with rice. She lived on soup and water. "I don't want to do it again. I need discipline and focus, because I don't have a lot of money."

When Kate returned the following year, Brandon was attending the San Pedro Roman Catholic School. In October 2000 Kate signed on with the Roman Catholic School. "Right now I have been with that school 6½ years. In future I plan to stay here more."

In 2000, Kate stayed at Martha's Hotel. On Monday of her first week on the island she went to the Roman Catholic School, where she introduced herself to the principal, Miss Roxanne Kay. Kate explained that she was going to be in San Pedro for six months and that she would like to volunteer in a class every morning. The principal said, "I'll introduce you to a teacher with beginners. That is where the need is greatest, where teachers need the most help." Miss Kay introduced Kate to Raquel Flores, a new teacher in her twenties.

"She explained that I offered volunteer service every morning. Raquel's eyes widened and a big radiant smile blossomed on her face. 'I had been praying for this,' she remarked. The sheer number of kids she had was overwhelming. Her sister was a Standard 4 teacher. I observed Raquel's class. The children were so lovely. I was helping her, correcting, singing and playing with the delightful children; I was absolutely smitten by them. Martha's Hotel was only two blocks away. I had a bicycle without brakes to reach school."

AT MARTHA'S HOTEL 2000

Kate paid Blz $550 (US $275) a month rent. She had a hot plate, a small fridge and a basin. She was making chocolate that she sold at the school. "I managed."

After Hurricane Keith hit in 2000, Kate helped to repaint Martha's Hotel to assist the family she had known since 1983. At Christmas the hotel was fully booked.

Sleep was a problem at Martha's, due to its location in the center of town. Because of the hurricane, many cables and wires had to be reinstalled, often at odd hours: power lines under the street, phone lines, cable TV. The walls were like cardboard, and the Jaguar nightclub next door partied until 3 in the morning every single night.

APARTMENT 2001

Elsa Paz, a neighbor, built several apartments before she was Mayor of San Pedro.

"'Will you hold me an apartment if I leave you a deposit? I want to live here in October,' I asked and left her 3 months' deposit before going back to Canada. In November 2001 when I returned at Halloween, I had lots of school supplies with me, along with library books from the librarian's discard shelf. A child's computer, an electric pencil sharpener, gifts, little things children would like, stickers, sparklers.

"My friend Elia sold them in her shop. She sells school supplies, textbooks, and gifts. In 1983 they had the largest store on the island. They sold everything from fishhooks to fabrics.

They supplied everybody with everything.

"Elia is a clever woman. She makes costumes for the dancers, wedding dresses, wedding cakes, children's uniforms, Christmas costumes, Halloween costumes. If a child says, 'I want to be a bat, or a butterfly,' Elia will make a costume."

SAN PEDRO 2002 – MOTHER'S VISIT

In March 2002 Kate's mother arrived in San Pedro with Kate's nephew. She stayed with Kate for three months. One sunny day, she went parasailing, no charge. It was her 80th birthday. She took pictures. Miss Elvi invited them to her restaurant for a free meal.

Kate was living in Elsa's apartment, paying Blz $600 per month. From 1-4 in the afternoon, Kate left the apartment because it was too hot. Often she would go to Sam's place and sit on the terrace. Her mother loved to go swimming.

"My mother knows I have this passion for educating children. Canada is so different. When I was 10 years at home with the children and only doing stand-by teaching, I often felt terribly lonely. I was not interacting with other people. Here I am interacting with many people every day. At home in a village in Canada it is hard to have your social needs met. Here it is stimulating."

BELIZE 2002/'03 – APARTMENT

Kate came back to Belize on November 1, 2002 and stayed again in an apartment with Elsa Paz. In March 2003 Elsa Paz became Mayor of San Pedro. Kate helped her with her English speeches. "I gave her some ideas for her speeches. This island sustains the whole nation. They provide Belmopan with millions and they get only a token amount back. Elsa Paz is honest and transparent."

FREE MUSIC SCHOOL

In Belize Kate misses a few things; namely, a variety of reading material, Canadian radio and her piano. She brought a keyboard to San Pedro, along with 150 pieces of music, "quiet romantic dinner music."

First she played at Elvi's Kitchen on Tuesdays, for Blz $40 (US $20) and a meal, at Lilly's Restaurant on Fridays, and on Sundays at Fido's under the same conditions. "I had troubles. The music was not appropriate. So I stopped playing there, and at Fido's. Bartenders wanted disco music. I didn't ask for work there."

Kate signed on as a teacher with Floyd's Free Music School. The school has a small choir. She taught them Disney tunes and Christmas tunes, and songs from The King and I. "There was a Christmas Party, eggnog, refreshments, a concert, and a full house. There was also an Easter concert."

BELIZE IN 2004 – MATERNITY LEAVE AND RESIDENCY

In 2004, a Canadian teacher took maternity leave at the Island Academy, and Kate agreed to teach in her place. Kate stayed in Belize for a total of 18 months, and gained 'Residency' status.

ISLAND ACADEMY 2005

"At the beginning of March the Island Academy offered me a three months' position. I accepted. I am grateful for the experience. I worked there March, April and May and received a weekly salary of Blz $600 (US $300). There is no school in June. It was a refreshing change: air-conditioned classrooms, only 11 students. No computers – this I found alarming. They wanted to keep me there, but I did it for three months only. My heart is with the needy children. In May 2005, I played the piano for the graduation and then I left."

Kate returned to the Roman Catholic school.

"My time is spent one on one with the students who are struggling. Pity is not an emotion that I indulge. Poverty is a state of mind. I am not a bleeding heart. Even if they are poor, poor, poor, their eyes are shining just as much. I help with the homework. Some children come from dysfunctional families, children may be neglected. On them I put my emphasis, or on those who have behavioral problems. I check the work, encourage students with a star. They can all be superstars. After 6 years I know most of the students. I want to be at the grassroots level. I am a guest in this country. I don't push, I observe, I try not to judge."

BELIZE 2006/'07

Two days a week Kate was playing the piano in an art gallery. A lady stood for hours in the doorway, bored.

"If she did something, made something, people would look, would watch. She could be productive. She was never given this impulse to create. The educational system is accountable for that. You get a lot of fulfillment from creating. If only the school system did something in this direction. Tourism is the key word. The education system here should be designed for that, preparing these students for a future in tourism.

"Let those children express themselves creatively. All that they learn instead is ridiculous irrelevant drills for the next exam, not something to prepare them for life. This is a failure in the system. Children are not encouraged to communicate, engage, verbalize with tourists. It is a missed opportunity. And if the children don't get interested and sparked in those years, they will become adults without a creative outlet. To produce something with their own hands is very gratifying, or to learn an instrument, to pick up a paint brush and draw. There is no pottery here, no clay.

"This is why I am teaching Belizeans to play music. I make Blz $50 and a meal for playing one evening at a restaurant. Fido's has a contract for musicians. Casper is a musician, he is Panamanian, a bass guitar player who plays with two guitarists, one drummer, and guests. One of the original seven is there and his name is Dale. His father Dale married a Belizean and is also a musician. His mother sells ceviches [conch marinated in lime, onion, chili, and cilantro] and nachos. Ceviche is the 'Belizean Viagra.' They make it in huge pots. It is what I would call 'rocket fuel.'"

WHY BELIZE?

"I wanted to get away from the cold. I am a swimmer, I live for the summer. I was a swim instructor; it was my first job. My house in a village in Canada is drafty; it is an old house, not insulated. I left because of that and because I love to travel. My needs are filled here. My major motive is to avoid the winter, appreciate and enjoy this culture. I do enjoy tourists. I keep a journal."

Kate moved on from Elsa Paz's apartment to Sam's Hotel on January 1, 2006.

"I like to live in a hotel, associate with people, learn, socialize. I love hotel life."

TIME TO BE FREE

"I have never been free. Now this is a free chapter in my life. I don't have a family commitment other than during the summer. It is a blessing to have their understanding and acceptance of this chosen lifestyle."

She writes letters to her family. Everyone in her community in Canada is interested, fascinated by her experience.

"They know I thrive on this. I have given up my car, television. I tell them how wonderful these people are, how lovely the children are. People are more warm and affectionate here."

Kate's daughters live in London and her sister lives in Australia.

"I want to be a little old lady here. In 2008 I will get my pension, then finances will be secure. Part-time work is stimulating, and if you are around young people you stay young."

Kate loves to wear a fresh flower on her ear that she picks from a tree. There is a Ziracote tree next to her apartment. "The flower reminds me to just stop and take the time to appreciate a blossom and not to take this experience for granted."

DEVELOPMENTS

"Many people come here with intent of staying, but then they find it difficult. Their expectations are too optimistic and sometimes unrealistic. Many people get involved in the restaurant business, but there are too many restaurants already."

Kate remembers a massage therapist who left. Her husband is Belizean. When she arrived in 2002, there were three massage therapists; now there are 14. Her husband, who was in Washington, sent her big containers of special oils, health extracts and supplements, and health foods. She initiated the residency process, but she didn't have the stamina to pursue it to its conclusion.

"She had learned a lot from a herbalist from a rural place on the mainland. She learned from this woman, a bush woman, spent time learning about the plants, how to harvest, how to prepare. She knew a lot about body, massage, about spirit, health."

KATE'S ADVICE ON LIVING IN BELIZE

"People come here with false expectations, liquidate all their assets, burn their bridges, then they find it is a struggle, because there are already a lot of restaurants and hotels, and it is harder for a business to survive. Very often these people are running away from something. But it could also be the cold weather. And people who have experienced too much stress. For people who just want to simplify their lives, this is a very attractive alternative. It is also deceptive; this is an illusion for these horizon watchers.

"People with good work ethic, people who want to do something for the economy, who want to work with Belizeans, they are often successful. There are some fine businesses here and they are well accepted. Other people with business ideas come here, but if too aggressive or if they underestimate Belizeans, they don't do as well. Belizeans have grace and they have an instinct for people. If they are not compatible, if their way of thinking is not compatible with aggressive newcomers, they react indifferently. They deal with anyone with grace. I have never seen a Belizean lose his or her temper."

BUSINESS OPPORTUNITIES

There were few tourist shops in 1983 and 1984, and no industry other than fishing and tourism.

"It would be an ideal location to introduce a co-op. There could be fabric produced, they could learn to weave. I have never seen a loom here. They get fabric from Guatemala. There is no fabric production in Belize. Women could work at home, organized into a co-op, fabric, weaving, sewing; this would be encouraging for women. Women could get a wage for working from home. I'd love to see batik here, fabric, and wax dye, outdoors activities. There is an opportunity for a home industry in weaving, and needlework – weaving mats, flooring. They could make paper from coconut fiber. I can see these types of things selling. They could make batik bikinis, sarongs, and hand-painted dresses. They could be making sandals, hats. There would be a market for hand-woven sun hats, palm tree baskets and trays. Children, 10, 11, 12 years old can learn these skills to pass on from generation to generation.

"I would love to see a creative co-op for women, see a facilitator, coordinator, to get women in their homes productive, raising their family and earning an income, producing Caye lime pie and cheese cake. Fabric, jewelry, massage techniques, serving the tourist economy. There is a lot of serving capacity to the first world citizens.

"This island is out of balance without tourist dollars. I wish there was more here. Now there is no fishing co-op anymore on the island."

FUTURE

"It will be interesting to see what the island will be like in 20 years. They say that there will be a bridge to Mexico. The idea fills me with dread.

"When the family arrived for the first time in 1983, there was no airstrip, no jet, no cruise ships came to Belize City.

"Musa, Belize's Prime Minister, said on the radio that there is a big extension underway in Belize, lengthening runways in Belize City, for direct flights from Europe, Canada, and Asia. He said this was a major effort to attract further tourism. This will be a major upgrade. Airlines will be looking at this as a destination.

"Construction here is escalating. After the hurricane people started to rebuild. There are huge resorts north and south of town and most of the coastline has been sold. A 70 x 140 foot beach lot sells for more than US $200,000.

"In 1983 there was nothing on the northern part of Ambergris Caye. The town stopped at the cemetery. We were encouraged to invest at that time. It is an ethical issue; I didn't want to buy from a developer. I don't want property. I prefer to rent from a Belizean family. I have no intention of buying or building or being out of town. I don't want to commute by golf cart or boat; in the rainy season this is a hassle. I would like to explore the Belizean mainland more.

"There are 900 students in the Roman Catholic school. I need that school population, to sell my chocolate. It will help with my daily expenses. This island has the reputation with other Belizeans of being fast, sinful. I would like to experience other villages. If you travel around you get a better understanding, more appreciation. But I must say over the 23 years I've been back and forth, the friendships with the families are strong, I feel a loyalty. Those people who have been successful here are those who give back.

"Once a house was burning, opposite Holiday Hotel, with a shop below, and a Canadian was living in an apartment above. The cause was electrical wiring. The residents were not at home when the fire broke out. The truck from the fire brigade couldn't pump the water out of the tank. So the Belizean community made a bucket brigade and used everything that would hold water. People formed a line from the beach to the fire and put the fire out by hand. A bartender risked his life to extract the tanks from below the house.

"The following Sunday, all day, they had a barbecue in Central Park, lotteries, people donated bicycles for an auction, even a building lot, tickets, food, they raised $70,000 (US $35,000) for these homeless people.

"This is a little island with a big heart. The response to difficulty is really heart-warming. These people respond in an emergency; they are loving, and giving to anybody in a difficult situation. Businesspeople here, newcomers, donated tremendous amounts of goods and services for this fundraising. A very good dynamic, and the house was rebuilt. Accommodation above, shops below. The San Pedranos are very generous with each other, very hospitable.

"One close friend commented, 'You are a light in the Community,' and my heart was touched…"

RETIRING IN BELIZE

"If people think of retiring down here I would advise them to look everywhere in Belize, travel extensively, check out Corozal, San Ignacio, the south. Enjoy the culture, bring your savings, the more time you spend exploring the country before committing, the better. Some people are water based; other people may want to try farming, the mainland, a slower pace. Take your time, stay open, travel, explore, and meet Belizeans. It is such an interesting cultural mix here, Hispanic, Carib, Garifuna, Mestizo, Mayan, Mennonite, and Creole.

"I recommend just taking time, take a year, see everything, see people, talk with people, it is intense. You pay a Blz $50 per month immigration fee for the first 6 months, then $100 for monthly extensions after 6 months. But you can get residency or come here under the Qualified Retired Persons (QRP) program."

PART V, CHAPTER 31: KIDS, KINESIOLOGY, CARBOHYDRATES AND CAYE CAULKER

Ernie grew up on a farm. He was born in Burlington, Ontario, Canada, on October 20, 1963. He has two older brothers and an older sister.

"We had mixed farming. My father grew crops: wheat, corn, hay, and we also had 100 acres of evergreen trees. Our farm was right beside the Six Nations Indian Reserve. It was very rural, 5.7 miles from town, a small town called Caledonia."

As an infant, Ernie had many problems with his ears, and from grade 4 through grade 9 he had many infections, forcing him to go to the hospital every summer to get a skin graft to rebuild his ear drums. The irony was that he loved swimming.

"I was an avid swimmer and in grade 8, at about 13, 14, I moved away from my family to a big city to train for swimming. I would swim two hours before school and two hours after school every day. I won a number of Ontario championships, and from the ages of 13 to 22, I went to a training camp every Christmas break with the university students, and then in March with the Ontario team, to compete in Florida."

During this time, Ernie discovered that he also loved water polo. At university, when he was forced to decide between swimming and water polo, Ernie chose swimming.

"I was a medalist in our Canadian University Championships, and in Hamilton at McMaster University I completed my Bachelor of Kinesiology (study of movement)."

WORKING AND TRAVELING

After graduating from university, he went north to Wasaga Beach to run a water park for the summer. His goal was to make enough money to visit Europe and especially his relatives in former Czechoslovakia.

"My father came to Ontario in the early fifties, where he met my mother in Toronto. She is Canadian. I went with my father to former Czechoslovakia. My father is from Chomutov. We have relatives in Prague, other relatives in Germany, other relatives in France. I wandered all the way through Europe, Czechoslovakia, then Frankfurt, Germany; there I left my father and went north through the Netherlands up to Norway to visit a university friend in Stevanga. Then I wanted to go to Rome. That was in fall 1986. My love for travel came from all the travel I did as a child when swimming. Every month we went to a different city in Canada or the US. My mother frequently went with me. That is how I began to love to see other places."

When Ernie arrived in Rome, the Pope had just returned to the Vatican, so everything was closed. Ernie went to Naples instead. From there he traveled to Pompeii.

"I remember in geography class in high school we had a play on this topic, the erupting Vesuvius, and all the lava that fell on Pompeii. I wanted to see it."

Next he went to Greece, making a short stop at Corfu before visiting the mainland. He hoped to climb Mount Olympus, to see the Greek gods, the originators of the Olympics.

"In kinesiology we did a lot about where sports came from. After being on the bus for four hours we had passed Mount Olympus and we ended up in Thessalonica. And there I met some travelers who were going to Istanbul and I was truly amazed. The masks were beautiful, the saunas were heaven sent, the people were incredibly polite and helpful. I spent many days just talking to people because they wanted to talk in English. And Turkey was not a place I had thought about. It didn't exist for me. It was the first place I was blown away by. I was just amazed."

Ernie returned to Frankfurt on a 40-hour bus ride to catch his flight home. He had been in Europe for 3½ months, and he arrived back home before Christmas.

"When I returned to the family farm, I stayed for not even a whole day and I decided to move to a big city, Toronto, and live with my brother downtown. With no job and a degree I found work in a convenience store. Then I found work, another job, working as a towel boy; I gave out towels and juice for hotel guests."

WATER, FITNESS, SOCIAL WORK

When Ernie attended university he worked as a lifeguard, both at school and in the summer at a water park. The water park had water slides and a wave pool.

"So that shows that all my life water was important for me. That's a big reason why I chose to live on an island later. Water is very calming for me."

Ernie found work at the West End YMCA as a fitness instructor in 1987. In the summer of 1988 he worked with juvenile offenders on an island called Wapus in Ontario, near Picton.

"This began my career in social work. In the fall I began to work full time with young offenders as a group home parent. I continued to do social work but moved into a high school in an open custody classroom for young offenders. After two years as a classroom councilor I decided to become a teacher and returned to the University of Toronto to obtain my education degree. I'm licensed to teach grades 4 to 10, ages 9 to 15. These are the ages of the young offenders that I had helped previously in social work and I believe this may be the last chance for young people to gain skills and insight for later life success."

Ernie graduated after one year, in 1997 or 1998, and immediately obtained employment with the Kawartha Pine Ridge School Board. He took 10 troubled boys from a senior public school, where they had been running up and down the halls disturbing all the classes and teachers, to the local high school so they would no longer create havoc.

"We went to the high school to remain in one classroom; they learned needed skills in order to gain entry into high school the next year. That was fun, very rewarding and an innovative new program for another rural community."

Then Ernie was recruited to teach a life skills program to Special Needs students in another senior public school. There he taught basic living skills including numeracy, literacy, cooking and character development.

CAYE CAULKER

"In 2003 I received a leave of absence from my school board and moved to Caye Caulker."

How did Ernie find out about Belize?

"In 1990 I came to Belize on the advice of a friend who had come here to take care of an elephant who was depressed. This woman came down here to take care of this elephant for eight months. It was somebody's pet. She came back to Canada and said that Belize was a beautiful country, still virgin and unspoiled. My partner and I toured throughout the mainland looking for an ecotourist area and we realized that we would need a lot more money than we had.

"Our return flight was delayed and I met Judith from Caye Caulker, who enticed us into returning that summer to visit the Cayes. When we did, we were hooked. We moved to Caye Caulker. It took us almost a year, but we bought a home in Caye Caulker.

"Numerous times throughout teaching students with special needs, I was ready to hit the roof. Then I thought of our home in Caye Caulker whenever I was stressed. I would close my eyes and I could see the reef and the waves washing over it. Instantly I was calm and ready to deal with the next problem.

"So my partner and I wanted to live somewhere different while our health was good and our backs were strong. In 2001 we sold a yearling, a thoroughbred horse, in the Canadian sales at Woodbine."

In 2002 Ernie and his partner sold another yearling with a disastrous result.

"We received almost nothing; we were in the red, in the bad in the sale. So we decided to give up horses and to go to Belize."

MOVING TO BELIZE

Ernie, Tim and Annalise own a house in Belize. They thought it was a fabulous buy. The lots ranged from US $50,000 to US $80,000 (tourist price not local price) and are 85 by 100 feet. They saw a concrete house for US $100,000. It needed lots of work. After 10 months of negotiations they finalized the deal.

"When I was a child there was a show on television called Gilligan's Island. It was about simple life and making do with what you had. Later in university we drew an island on our dormitory wall called Bubu's Island to escape the winter blues. And now I find myself realizing my dream of living on an island."

Ernie's partner spent two winters in Belize before their move, to vastly improve the structure of their home, which they bought in 1999. Ernie's partner is a baker and a horse trainer. They moved to Caye Caulker with their bird, a lesser sulfur kakadoo; their 15-year-old Jack Russell Terrier; and "a wealth of dreams."

A VOLUNTEER TEACHER

When they reached Caye Caulker, Ernie volunteered for four months at the San Pedro Catholic primary school.

"I was amazed at the number of hours devoted to religion without any apparent character improvement. I was appalled to see the lack of resources for students and teachers alike. But the students' need for attention and yearning for knowledge is no different from that of other students. They look for your approval as an educator.

"For the summer I helped a troubled youth in San Pedro gain insight into his unruly behavior while catching up on required skills. The following year I worked at the private school Island Academy and again was appalled at the lack of resources for teachers. Unlike the Catholic school, the private schools pay for and receive excellent textbooks, but no manipulative or higher order thinking is apparent or is in sight."

Traveling daily from Caye Caulker to San Pedro ate more than ⅓ of Ernie's salary.

"But a successful year was gained through the development

and growth of students in pursuit of getting their PSE (national exam). Even our special needs students received passing grades in the national exams."

In March 2005 Ernie returned home to visit his dying father, who had decided to forego another shunt site in order to continue dialysis. Ernie's dad had been on dialysis for a number of years and his quality of life had declined a great deal.

"I was fortunate to have lived a lesson I had just taught to my Belizean students, of honoring their parents while they are alive."

CADETS ORGANIZATION

During his first year and a half in Belize, Ernie worked with the Cadets' Organization, which is a police youth organization based on sports and cultural activities. The first camp he went to for cadets was in Benque Vejo, right on the border with Guatemala. They went there for camp and Ernie was asked to sleep with 30 other children in a classroom on a concrete floor. Ernie borrowed a gymnastics crash mat so he could walk the next day. But 30 kids had to sleep on concrete.

"A fundraising barbeque put on by the cadets raised Blz $1500. Mysteriously when we arrived at the camp we only had $500 to purchase needed water and food items for the campus. When we got there we said, 'Let us have some money.' They told us, 'We only have 500.' The police chief only gave me 500 and they kept the rest.

"That being said, the police do not make enough money

here, and it leads to corruption and stealing. In Caye Caulker it is a running joke, police come with one empty duffle bag and leave with three full suitcases. People laugh about it all the time, but it happens. The police don't make enough and that's what leads to these problems."

The Cadets' program is modeled after the Boy Scouts program. It helps immensely for the youth to see all of Belize and understand other areas more clearly, because the cadets who went from Caye Caulker to Benque had never been to the mountains in Belize.

"It is a wonderful program, but... the nutrition in that camp was poor. No daily recommended food standards were met. They didn't get enough proteins, carbohydrates, and, of course, vegetables. That was a big concern.

"We had for one meal corned beef hash and bread. The corned beef was mixed with hot water and it looked like gruel. My campus couldn't even smell that. So I took them into town to get chicken. That was one example. You have proteins, water, bread, no fruit, no vegetables, not enough calories for a growing child."

A LONG TEACHING DAY

Ernie was up at 6 am, he took the boat at 7 am, and he enjoyed breakfast with the Liars' Club at Ruby's. The Liars' Club is a group of expatriates who get together to swap stories over a cup of coffee. And that became important later on, as Ernie was introduced to his present employer at the Liars' Club. "That is how I became a real estate agent."

At 8:00 Ernie was at school. Classes began at 8:30 and ran until 3 pm. "I also taught PE from 3 until 4 for different classes for each day of the week. I taught Standard 5 and 6, all subjects except Spanish; Mr. Will taught Spanish. And then at 4:30 I caught the last boat home to grade papers and prepare for the next day."

LOSING HIS HEARING

During the Christmas Concert in 2004 Ernie noticed that he couldn't hear the music or the actors. He realized he wasn't hearing.

"My neighbor, a nurse practitioner from Canada's north, confirmed both of my eardrums had holes. I tried in vain to help them heal and my left ear healed with medication. I decided I needed to take care of my ears when I still didn't hear well in September 2005. In Belize City I went to a specialist. It wasn't until November 2005 that I went to Guatemala to have my ear operated on with great success by a Baylor-educated specialist.

"A Spanish-speaking friend accompanied me from Caye Caulker to Guatemala. In Flores, while I was searching for medication in the form of eardrops rather than pills, an otologist was having a clinic and asked what my problem was right in the pharmacy. I was so lucky." This otologist operated on Ernie later.

Ernie had been out of Canada for more than six months, so his medical insurance was invalid. In Belize they wanted US $5000 to do a skin graft on his ear. In Guatemala they asked for US $2000 for an experienced surgeon and a non-invasive European surgery. They sandwiched biomedical skin and Ernie's skin to prevent rejection.

"There are a great number of scars on my eardrums from childhood surgeries from the early '70s. They did this ear surgery in Guatemala City. The views from the operating room and the recovery room were like three to five volcanoes on the horizon, accompanied with waking up underneath three heated lamps. I was rejuvenated in spirit, it helped the recovery, it was incredible, it was amazing.

"The surgery was a success. After three weeks of healing in Antigua, I returned to Caye Caulker to finish tutoring a student in physics back in San Pedro."

REAL ESTATE

"When I reached San Pedro in December 2005 I was approached by a member of the Liars' Club to work for him in his real estate company. Every Monday to Friday morning, a bunch of expats get together for coffee and swap stories at Ruby's Hotel. People talk and solve many problems. And there have certainly been some moments of frustration and high emotions amongst members regarding political situations here in Belize and abroad. Americans and Canadians, well-educated people and well-experienced, are all willing to give to you, to share their experiences. And that is what the Liars' Club is, predominantly men but there are also women."

There Ernie met Mickey. He offered Ernie a position with his real estate company, which he had just opened in June

2005. Ernie continued to tutor students in San Pedro in preparation for the PSE and to tutor needy students in Caye Caulker.

"I am a real estate agent. I am trying to sell homes, match clients with sellers. I have to find properties and in February 2006 I opened another real estate office in Caye Caulker for the same company. The company is Belize Shores Realty, selling in San Pedro and Caye Caulker."

LIFE IN BELIZE

The other day when Natasha, a friend from Caye Caulker, was in San Pedro she talked about Ernie being brave.

"I never thought about that but she is correct. Belizeans are homophobic. Some people accept it, some people don't talk to us. In fact teachers in Caye Caulker asked my neighbors what type of person I was, for fear that my sexuality would get in the way of teaching students. And fortunately my neighbors understand that sex has nothing to do with either being a teacher or being a good person."

Ernie is a real estate agent on commission. One sale equals half a year's salary as a teacher. "I think it will pay for itself. I came to Caye Caulker to find out more about myself and the world. I have always tried to be in the safe zone, and this was an opportunity. I am working with Bob, a team player. The owner is a stereotypical American. He has few social graces, is easily angered and we work 100% commission. I am fortunate they pay my water taxi here and home."

In Caye Caulker Ernie and his partner have a garden with their house. "We grow tomatoes, okra, basil, chives, we make our own bread, and we never go to eat in a restaurant. I do tutoring, earn 400 dollars a month, and live on it.

"Here in Belize they pay their teachers nothing and expect to have a world-class education; they don't remunerate their teachers. We have teachers in Caye Caulker who raffle cupcakes. They make $50 from the raffle and the teacher takes $30 for making the cakes. They have kids they have to take care of too. The government provides for salaries, not for any books. There are some donations to the school.

"There is a computer lab at the Caye Caulker School I helped to re-establish. I tried to organize the teachers to learn about computers, but no one showed up and I think it is because the administration is afraid to admit that they don't and can't use a computer and therefore the students lose. Because to this very day the kids don't use the computer lab, which has ten donated completely new computers and that is sad, because this is an opportunity for students with deficits in any area to receive immediate feedback and constant practice of skills yet to be mastered. And the teachers don't know the value of using a computer.

"At home in Canada the first thing I do with my Special Needs students is to introduce them to a computer. Many have skills, learned by playing computer games, and they find immediate success with computer programs.

"The director of the school in Caye Caulker is technophobic. She doesn't know what value she has with the computer lab with ten computers. It is sad. There is a woman working in Caye Caulker, Ellen Armstrong. She runs the Keyboard Connection, an Internet café, and she runs a raffle about every six weeks with products donated by businesses. She's raised over Blz $2000 for the computer lab.

"And she is continuing to do that. She sees the need for it. She's trying to get five more computers, money for maintenance and staff training. She's just raising money. She receives no help from teaching staff, but ironically she has recently been asked for money from her computer fund to fix the photocopier. She said, 'Yes, but the staff must earn the money first. If you want to sell some tickets in the raffle, you can use some money for the photocopier. You need to sell some tickets.'

"She wants them to be interested in this fundraising for technology. And she will get there. Her child is just in Standard 2 now and she knows the need to use computers to keep up with the world."

Ernie also works as a correspondent for the San Pedro Sun.

"I just wrote an article on her. I am trying to get more Caye Caulker information into the paper. In January they asked. I taught their son in Island Academy, and I teach four high school students over the Internet. I'm a facilitator. They're linked with the Keystone Program in Pennsylvania. It's an American high school online. It's very expensive. The families pay for it.

"The standard here is not the standard of an American high school. The depth of knowledge, the mastery of skills and the consistent practice is lacking in the Belizean high school curriculum, along with the lack of resources. There isn't even an overhead projector in the private school."

BUSINESS OPPORTUNITIES

"Any business opportunity needs to be linked to the tourist industry, because that is the hard cash coming into the country. I think water will play a vital role in development anywhere in Belize. A sustainable agriculture is an opportunity, including the development of hemp (the non-medicinal marijuana that doesn't have THC in it), because you can use it to make rope, to make cloth or textiles. You can get oil that is better than any other oil, nutritionally, and there is a product called hemp nuts, also another nutritional supplement. This climate is perfect for growing hemp. They are starting to grow hemp, starting development.

"Like with cotton. There is a little bit of initial investment of setup cost for machines to process a product into a finished product but it will reduce their need for imported materials and manmade polluting plastics, like nylon rope, nylon bags for sugar, flour. They could make these bags and reuse them. They will get there, it just takes acres and acres of it."

ERNIE RECOMMENDS:

If you have a pet bring it; it will keep you sane.

Leave your cookbook at home and bring a very open mind.

Always be responsible for yourself first.

Don't worry about money, it is a trap. Many people think that you need to have a lot of money, but you don't. It may even shelter you from some life lessons. This applies to younger people, not to people who are retiring. If you have too much money you continue living as a tourist and you aren't really living here, you are just visiting.

Listen to everyone and be careful when you speak out, because you can easily offend someone's culture.

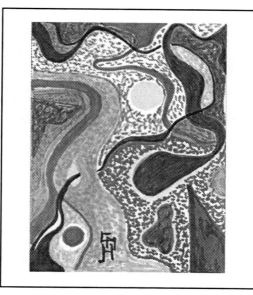

Organic Shapes...

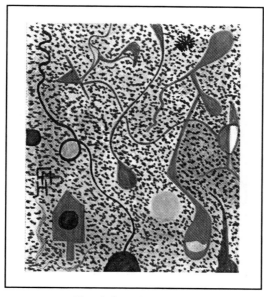

Explode in Vibrant...

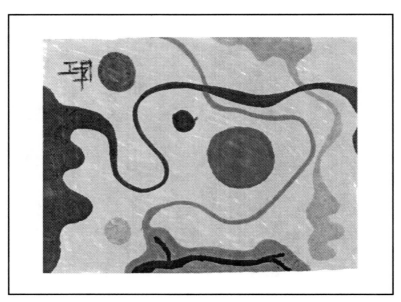

Joyful Colors...

With a Smile of Paul Klee

The Play of...

Vital Forms.

PART VI: RETIREES

PART VI, CHAPTER 32:
A POLICE OFFICER RETIRES

"I first thought about going to Belize in 1999. I was on the point of retiring and was looking for a place to live. I'm from the US. I wanted to leave because of high taxes, snow, most of all the snow. I wanted to be somewhere else, to write, but I haven't written. I wanted to write novels."

Leslie Couillard was born and raised in New York City. "I had a lot of jobs before I went into the military in 1963-70." Before retiring, he was a patrol officer in New Jersey. At other times, he worked at the university library and he was a prison guard. He ran his own mail order business before he finished high school, selling stamps to collectors.

AN EVENTFUL PAST

"I went to a lot of places. I was in New Jersey for training, in Korea after the war, in a hospital, in the medical unit there. I was assigned, went to Utah, then to Korea for one year. I had been trained as an operating room technician, surgical technician; I did that in Korea, in Utah emergency, later surgery. In Utah there was testing of chemical biological and radiological weapons, but I was working in the hospital."

After Utah, Leslie was stationed at a hospital in Verdun, France. "My commander knew that I wanted to go to Officer Candidate School. So I was sent to the non-commissioned officer training academy in Bad Tölz, Germany, for six weeks."

Leslie trained in Fort Benning, Georgia, for six months to be an officer. Then he went to Fort Hood, Texas, where he was in the mechanized infantry. "You ride around in armored vehicles. From there to special warfare school, and from there I went to Vietnam."

Leslie was in Vietnam from May 1968 until December 1969. These were hard times, but he enjoyed the friendly people and warm climate, although sometimes it rained. Later Leslie looked for a tropical climate to retire in.

"I even thought about going back there. I worked with the Vietnamese. I wasn't with the Americans, I was an advisor to the Vietnamese. I worked with them, ate with them, lived with them. I was in a Military Assistance Command Vietnam, with the South Vietnamese. I was with a team of about seven that was assigned to the Vietnamese District Government. I have a different view of the history of that time than everybody else."

Leslie returned to Fort Dix, New Jersey, to serve the rest of his enlistment. He finished in May 1970.

"From there I was a corrections officer, a corrections sergeant, a corrections lieutenant, then I left after ten years, 1979/'80, that was when I left. Then I spent a few years working in a college library, and for the welfare department, and then returned to correctional work as a patrol officer in New Jersey. The patrol officer is outside in the community, not inside the prison. Before I was inside the prisons. They used to call us prison guards. I remained a patrol officer until I retired in 1999."

Leslie switched to another department for his last two years.

New Jersey had created a new program for juveniles. Everything having to do with juvenile offenders was to be placed under one commission, the Juvenile Justice Commission. They took over the juvenile prison and other things that had to do with juveniles. They wanted to have patrols. "I went over to the Juvenile Justice Commission to teach them how to do it."

RETIREMENT

When Leslie thought about how he wanted to spend his retirement, he remembered how much he had liked the climate and the people in Vietnam. "I looked at Central America, South America and Mexico. But I didn't want to rely on my high school Spanish. I even considered Vietnam, Thailand, Cambodia, Philippines, searched on the Internet.

"One day following an offshore opportunities' link I discovered Belize. English language is spoken, it has a stable democratic government, and here I am. It turned out to be just what I was looking for. I just got on a plane and came."

He knew a lovely lady with three children. She left them with their father. She wasn't able to work on the island and had no friends there. Then there were troubles with her children back home. She left after three months and hasn't come back.

Leslie's first marriage was to a Puerto Rican girl, back when he was selling stamps. "I learned about rice and beans; that is why I am also happy here, rice and beans." His second marriage was to a Korean nurse in Utah. He has no children.

Leslie lives in a nice little one-bedroom apartment near mini-golf.

FLASHBACK

"When the plane landed it felt like Vietnam. Painted signs all over, houses made of wood, cement, some made of plywood, aluminum and cardboard, like in Vietnam, it looked like that from the air. This was in August 2001. When I looked out of the balcony or window of Martha's Hotel I could see an empty lot where the Barrier Reef Hotel was before they tore it down. Now the Alliance Bank is there. One of the many changes in the five years I have been here."

SETTLING IN

It was relatively easy to get settled, but not totally without problems. For example, Leslie had to go to Belize City to withdraw money from ATMs because there were no ATMs in San Pedro.

A friend came to visit in the first month after Leslie's arrival. They rented a golf cart and went north across with the ferry. Leslie felt frightened. They picked up four or five kids. "It took me and all the kids to beat off the mosquitoes. We got stuck in mud in the streets up north. Now the north is a pleasure to go up to, beautiful, but at that time, there was not much there.

"When I first came, I did all my work in the Internet café,

the CocoNet. Peter Englisham used to own it. At lunchtime I used to lay in the sun. From 1-4 the Internet café was closed. San Pedro Fitness Club has a pool, the largest pool on the island. At the yacht club, there are also fitness facilities and a pool. I work out. The girls in the family come and get me several days a week to take me to the fitness club."

LOGGING ON

When Leslie was in the police academy, there was a terrible snowstorm in the winter of 1995/'06 and he couldn't get out of his house. The academy closed for a couple of weeks.

"I had nothing to do, so I plugged my computer into the telephone and dialed up AOL, got online and have been online ever since. You can earn money by being on the Internet.

"The fact is, I don't make any money, it costs a lot of money. For a three-month period I made very good money; now I don't. This is a combination of a lot of things. Most of my customers are rebuilding in New Orleans and Florida after the hurricanes. They haven't money to spend on the Internet; they spend money on other things.

"Retirement covers everything. I was making enough to be comfortable down here. I really do my business as a hobby. It gives me something to do."

A BELIZEAN LIFESTYLE

"I got down here, got my passport stamped, went to Belmopan after two years, and put in for Residency. It took me a long time, but now I've been here long enough to put in for citizenship in August [2006]. That really has been the plan all along.

"What I like most about staying here is the climate, the weather, the friendliness of the people. After knowing you for a few months they invite you home. That is what I like. Nice people, climate, beaches, beautiful places on the mainland, lots of places I have not yet been, nice clear water. And I love the food. I cook myself. Rice and beans, spaghetti sometimes, chicken, vegetables and rice. I have a housekeeper who cooks stewed beans. She puts peppers, whole, small ones in stewed beans. I pay Blz $150 a week, she comes three times a week for a few hours. She does shopping, everything. Does laundry, irons clothes. She's as reliable as a 15-year-old high school kid."

LESLIE'S SUGGESTIONS

"Bring money and bring skills down here if you want to come. If you're retired and don't want to work, nothing is wrong with the retirement program. If you have money and do not need to work, the retirement program is best.

"If you have money come down here and open a cannery. Belize has enough food growing now. Make money by exporting things, but also make cans for here in Belize.

"You can eat for way less than US $20 a day. I pay more eating at home than I would in the little restaurants.

"The other thing about San Pedro is, I stay a lot to myself. Local or expat community, this is a party island. Tourist or whoever, if you are a sociable type of person and want partying and drinking it is a good place. This has been described as a drinking island with a diving problem.

"Compared to the rest of Belize, it is expensive here, but much cheaper than most places in the States. My rent now is less than US $400 including free cable and electricity and water. I have a small AC in my bedroom and a big AC in the living room for the entire house.

"First of all you need to show US $1000 monthly income to get a stamp, or a ticket back to the States. You have to show that you have Blz $2000 per person per month. If you have that much money, that is enough, you can get by in San Pedro. You need considerably more to get by on the mainland US. So Belize is ideal for the retired on Social Security. I am a rich man for two days and a broke man for the rest of the month: restaurant, supermarket and clothes, accounts; once my check is here, I pay."

THE FUTURE OF BELIZE

"They need some traffic lights down here. So many taxis and trucks, it is becoming a small metropolis. Another town may be built in the south, a town up north, maybe three more towns on the island. It depends on the government. Supermarkets will be needed, and Internet cafés. There are four regular international banks: Belize Bank, Alliance Bank, Atlantic Bank, the former Barclay's Bank that's now Caribbean.

"It's just getting bigger and bigger. Now there are some cobblestone streets in town; there will be more.

"We are in for a lot of changes. The development depends on politics as well. I believe that Belize is going to be suffering from what the US views as democracy. The PUP [People's United Party] was elected, people were upset, the UDP [United Democratic Party] was elected here on the island. It takes time to achieve political objectives; 4, 6, 8 years may not be enough.

"A frequent change from PUP to UDP doesn't give either party an opportunity to accomplish its objectives. For example, in San Pedro the PUP-built park had shows. The UDP just tore it down. The basketball field and football field are no longer there. The football field is a parking lot.

"Real estate will be a good investment for the next few years. Those who have a lot of time can invest in some place in Belize that will become a tourist attraction tomorrow. Placentia, or any place on the beach, attracts tourists. Rain forests, resorts, Mayan village. Run tours from there. Diving tours, etc.

"Karin's husband built the first steel mill in Belize. He goes around to junkyards, gets iron from old cars, and makes rebar for houses. And he did it without financing. His wife has a clothing business. He went into the fertilizer business. He sells it to the States, and chicken feathers, to support building this steel mill. Now he has a steel mill, the first and only one in Belize.

"So someone could come down and make a cannery. I

would like to see things happening here besides the tourist industry. Canning would support the economy here. They would need workers. People here are working in the fields, and they are farmers. A lot of money can be made in tourism, near rivers, in the rain forests, ecotours, archaeological tours. With all the things going here they could build a university to study archaeology.

"Perla Escadido has condos on the beach side with a swimming pool and a management team. Beach side US $170,000, lagoon side US $120,000. It would be better to build a cannery factory instead of buying a condo. There are so many things we could use here besides condos.

"The people in San Pedro are very friendly. On the mainland they are even friendlier. I met a lady from Caye Caulker. She went to work for a diving company. She came over here and she said, 'These people are not friendly.' She thought they were very unfriendly because she had nothing other than Caye Caulker to compare them with.

"San Pedro people realize that half the population are not San Pedranos, but are from Central America, Mexico, USA. People here come from Guatemala, US, Canada, El Salvador, Honduras. Many come from mainland Belize, some from Germany, France, Holland, Belgium. The owners of a resort in the north are Belgian, Dutch. The point is it is a cosmopolitan community like New York City, where I come from."

PART VI, CHAPTER 33:
AT HOME IN BELIZE AND THE U.S.

Janet Smith (real name known to publishers) worked in broadcasting in the US. When she and her husband first arrived in Belize in 1989, she didn't think they would stay for so long.

"I assumed we'd spend just two years down here and then go back to the States. But my husband had a different view; he fell in love with Belize the moment he set foot in it. We came from the Miami area. Belize and Florida are similar. The currency was understandable; that made it a little easier."

Janet was born in New York and lived part of her younger life there, through high school and a portion of college, and she loved New York. "I wanted to be a doctor, then a lawyer, then I got interested in politics and a lot of subjects. When I got married the first time I was a virgin. A very foolish mistake."

A CAREER IN NEWS RADIO

At the end of the '70s and during the '80s Janet worked in news radio as an anchorwoman and an announcer. She hosted a radio show in New York called "The Action Woman."

It was a half-hour show in four portions. The first part was a woman in the news. Janet gave reports on women undertaking different affairs. Second was an interview. "I interviewed a number of different women: a girl that flew a plane, women who did marvelous things. I had the girl who wrote 'The Color Purple' [Alice Walker]. The majority interviewed were African American women, but also people from everywhere, from all societies; they also talked about events that occurred." The third segment was called 'This and That' and focused more on the humorous side. The final segment was 'Consumers' Corner.'

"This was black radio to a degree, once a week, in New York. In the '70s in Florida, all I did was news – at various stations in the country."

During her time working in radio she also interviewed Jimmy Carter and President Ronald Reagan. When Jimmy Carter once came to Florida, a news conference was scheduled at the Miami Herald Newspaper. Janet arrived late because of parking problems and nearly missed Carter, who at the time was running for president. The press conference was over when she suddenly found herself face to face with him. Janet was very excited.

"I stood up in front of him, I called him Reverend Carter, Dr. Carter, everything except President Carter. He took my hand and said, 'The correct title is President or Governor Carter.' I was hoping for his victory; he was nice."

When Janet wrote, she wrote in the plainest English, in comprehensible language, to be understood easily. She dislikes the sloppy style often used today. "People are using unnecessary words; they could speak clearer, more understandable."

Janet is quite knowledgeable about politics because of her

work in radio. "A lot of people don't realize that their constitutional rights are in danger now in the US. Now there are so many decisions to be made. I regret that I didn't go into political science. I wish I had gone into politics."

WHY BELIZE?

The main reason Janet is living in Belize is because her husband's father was born in the Stan Creek District of Belize. When her father-in-law moved to the US, he said he never wanted to go back to Belize. Janet thought of her father-in-law as an ignorant man.

"He bought clothes, suits, shoes; he became a Don Juan as far as his wardrobe was concerned. He couldn't stand Belize when he talked about it. We thought, 'Why don't we visit the country?'"

He died at the age of 89. After he died, and Janet's mother-in-law died, Janet's husband took care of the estate papers. Then he desperately needed a rest.

"We left New York to go to Miami for a few days for the sun and relaxation. Then my husband wanted to go to the Bahamas. He was ready to go at the airport when he called me. It was in 1988. He said, 'You know, I see Eastern Airlines flies to Belize. The sign is up here. I think I want to take a shot at it.' He bought a ticket to Belize instead. When he got to Belize, he booked a room at the Fort George Hotel in Belize City. He called me that evening and said, 'Pack up your clothes, this is the place where we're going to live.'"

A BAD FIRST IMPRESSION

They lived in Belmopan at first. "However, I was not that impressed."

Janet thought they would only spend a year or two in Belize, seeing her husband's family. But when he looked up his relatives and found them, a half sister and nieces and nephews, he fell deeply in love with his country, and after a few months he applied for his Belizean citizenship through ancestry and got it without any difficulties at all.

"I was so surprised. I was thrown into a lot of emotions and feelings. I love the US. It is my home. I wasn't thrilled. I liked it when I thought it was a short stay, not permanent; however, it turned out on my husband's side as permanent. He bought land in Camelotti, 20 acres undeveloped. We sold it a number of years later."

Her husband had a job with Maya Airlines. He had purchased that land in Cayo District, 20 acres for Blz $5,000 (US $2,500), almost 20 years ago. He went from Belmopan to Municipal Airport where he worked on aircraft. Janet was a housewife who stayed at home with her dog and her cat. In the evenings they went out for dinner. Because he worked for Maya Airlines they had free passage within Belize. They would fly down to Placencia.

SAN PEDRO

"Then San Pedro found us. And we liked the place; I liked it. One day we flew there. We became so familiar that we decided to make it our home. It was with the same airline, so he could work in San Pedro. We came out and built this little house in 1993, 1994. We stayed at Sam's Hotel, but not that long. Maria and Sam and their kids I knew well. We loved it here. I lived in a few places. It was livelier; we had no responsibilities, only a dog and a cat.

"Then we moved out to our house. Mr. Campbell built the house. My husband was thrilled with the guy's work; I was not. We paid cash for the house; not everyone did that. Now all the millionaires pay cash when they come into the country.

"San Pedro was lovely at that time. It was not as commercialized as it is now. Now many people have remarked that it seems that San Pedro is trying to compete with some of the Mexican resorts. A lot of the quaintness has left. It is not as pure as it seemed to be. I miss the quaintness. The people were very close to each other. With the commercialism a lot of the quaintness filtered.

"San Pedro still maintains its beauty and especially on the northern part of the island there is a great deal of raw land with bush. New resorts and hotels are constructed along the sandy beaches up north and on the lagoon side in the south.

"I saw the change. People have always been kind and friendly. I was the one who was dissatisfied. The basis of all of this is the heat. I was not used to being exposed to year-round heat.

"When we moved into the house I loved it. I still love San Pedro, but San Pedro doesn't love me. It is not the heat, it is the sun. I don't want to see a hurricane start up. The rainy season can be a little uncomfortable. San Pedro is overall a good place. I do like the island better than the mainland. A lot of tourists are from Canada and the US. The Belizean portion of my life is wonderful, but I just like to get home, just to get a taste of the States. I love my country but I never will forget Belize. I consider this here my home as well as the US. I have two homes now. I lived in Florida, New York and New Jersey. I have a few friends in the US. Many have died. I am one of the oldest on the island."

Janet and her husband had property in New Jersey that he had inherited from his family. They had two houses. Her husband had fallen so in love with Belize that he made up his mind to sell the houses in New Jersey. In February 1990 Janet got the power of attorney from her husband and sold the two homes. "Selling the houses was the biggest mistake we made." They paid cash for the house they built in Belize and used the money for traveling and expenses.

JANET'S THOUGHTS

"During these 20 years I have been living more here because I cannot afford the States. USA is home to me. But in Belize you get more for your dollars than in the US."

Janet sighs and goes back in time: "I believe all the testing, the playing around with, experimenting, trying out a bomb in the '50s — we are feeling the results now. The testing of the nuclear weapons takes 50 years; the Nevada desert is polluted. There are experimental laboratories. The disasters we have today are the result.

"On the spiritual side, man is playing with the elements, touching what he has no business touching. I think people have lost their feelings for other people. The human has become more like a robot, looking the same, dressing the same – a lot of sameness. I believe that there is a notion set up. People don't care about older people anymore. Your mother and father were the most important people in the family, but now I believe people today don't care about older people. The elders are really important; they hold the key to a good life."

ANOTHER INTERVIEW

A day after my first interview with Janet, she went to the United States again to stay with friends for a while. We met again after a few weeks when she returned to San Pedro. We continued the interview. I remember the last words of my interview with her in June. In her mind she returned to the days of her youth.

"I have a high school diploma. I attended City College of New York, but I had to give it up because of financial reasons. Spirit wasn't there. I wish I could do it all over again and cut some of the mistakes. I would love to lecture, impress them how important an education is.

"If I could live life over I would be in show business. I danced, I was a dancer too – waltz, foxtrot, rumba and tango, I loved it – I was not on a professional level. I probably would have made it if I had followed it through. Now I feel so incapacitated; now I cannot dance any more.

"I always worked. I worked when I was in high school, at a dry cleaner two hours a day in the afternoon. If it is something I want, the activity, the wheels begin really to move.

"I have done a lot of things. I love to dress up, I love shopping, I was a shopaholic, and I was hungry for clothes. I spent a lot of money foolishly on clothes. I regret it, but at that time I was happy. Money: go somewhere and dance some more."

ON THE BEACH

A few months later Janet suddenly passed away. Her husband's adult daughters, who she dearly loved as her own, came from Europe, and relatives and friends flew to San Pedro from the US, Canada and Europe. Many expatriates mourned with them. They lost a good friend. The local newspapers reported. Janet's husband sent a poem he had written for her.

His daughters, who loved Janet and who Janet loved, gave a farewell beach party in her memory. People came from Europe and the US and many friends from San Pedro attended. There was a large buffet and drinks for the many guests. Everybody told stories and anecdotes from Janet's life. Many of them were funny little tales and there was laughter on the sunny and sandy beach.

Janet would have loved to have been at this party which was a farewell to her in deep love.

PART VI, CHAPTER 34:
A PARTLY RETIRED LAWYER LIVES HIS DREAM LIFE

Pat Stiley started practicing law because his roommate, a conscientious objector, didn't want to join the US army in Vietnam from 1968-1972.

He began practicing law in 1968 as an intern and in 1971 as a full lawyer. In 1973, he quit, and from 1973 to 1976 he traveled, though his employers in Spokane, Washington, kept flying him back there to argue cases.

"I came to Belize, Nicaragua, Peru, looking for a place to settle. In 1976 I went back to the US to practice law. I was tired of not being able to afford cigarettes. I stayed 30 years and practiced law. I came down to Belize to visit, but I didn't retire."

When he returned to the US in 1976, Pat told the company that he would practice law if they made him a partner in the firm.

"1976 to 1999, 23 years, I represented nothing but little people. I refused to represent companies. I represented engine workers, people charged with crimes and civil liberties people. For the right to free speech, freedom of religion, the right not to have a Christian symbol in front of you when you are a Buddhist, the right to gather to promote unpopular causes: Communists, homosexuals, people who wanted to be naked on beaches, gypsies who were offended by being strip-searched.

"In 1999 I decided that technology would allow me to do my work from other places, since I had always traveled. I grew up traveling. My father was in the US military. My first language was Japanese. I first learned to talk from a Japanese housekeeper and a Japanese cook. My second language was English. From 1948 to 1950 we were in Hokkaido, Japan. My father was a captain. He had to hire three Japanese people in his house to help the economy. He had to hire a housekeeper who didn't keep house and a cook who didn't know to cook, but they taught me to speak."

Then they returned to the US. At that time Pat's father went to Korea. Pat lived in the US with his mother, his two older brothers and his older sister until 1953. Then they moved to Germany. His older brother and his older sister went to American schools there. Since Pat was the youngest, his mother decided to put him in a German grade school.

"I spent two years in German grade schools until I was dreaming in German. When I wanted to go on the base they wouldn't let me in because I spoke English with a German accent. I grew up traveling. So did my brothers and my sister."

Then the family went back to the US.

"In the 8th grade I realized that I was too young to be in Castro's revolution. I went to the seminary to be a Catholic priest, a missionary, a Franciscan priest. On Mondays we were only allowed to speak Greek. During lunch and dinner someone would read stories in Greek. On Tuesdays we spoke Latin, on Wednesdays we were only allowed to speak Spanish. During the meals they would read stories to us in Spanish. On Thursdays we were allowed to speak English, Fridays German,

Saturdays and Sundays English."

When Pat's Spanish language skills had improved, he started traveling to nearby Mexico and Latin America. Pat decided not to be a priest.

"If God really wanted me he would call later. But he didn't call back. I started traveling Latin America from the age of about 16. I skipped a lot of years of school. I started college at 16, was a lawyer at 23. I enjoyed partying. I enjoyed government guaranteed loans, poker games, and pretty girls. I have known for a long time that I would spend a large part of my life beside a warm blue sea. When I was in the 6th grade, when the nun said draw, I would always draw a blue-green ocean, a sailboat and pretty little white clouds over it, sometimes a little isle with a palm tree.

"In 1980, when I could afford to travel again on vacation, I started to go to tropical places and rent sailboats for a week or two. I did that in Asia, in Thailand, but I was getting too old to learn new languages. Thai was tough, so I learned Japanese again. From 1965 to 1980 whenever I went somewhere, I went to the West coast of Mexico to fish and drink and chase women."

He spent a few weeks sailing the West Indies. In 1993 a sailing companion said, 'Let's go to Belize." Instead of a monohull, for Belize we used a catamaran; it is not so deep a craft. I really liked that."

Pat had been married and divorced twice. He later remarried, to a woman with a six-year-old daughter.

"Around 95, 96, we rented a catamaran, right here in San

Pedro. My wife knew I would telecommute; I was going to live on the beach and practice law. She knew I was looking for where I wanted to be. The second time we went to TMM [a worldwide sailboat charter company with a base in San Pedro] in San Pedro she saw the Island Academy school and said that's where she wanted her daughter to study. Therefore we decided to move to San Pedro. I was thinking about Cuba or Grenada, but my wife picked San Pedro.

"At that time I had a law firm in the US and I was a pretty well-known criminal defense lawyer. I had five lawyers and god knows how many interns working for me. So I picked the lawyers that had been there longest and said, 'I have a good deal for you guys. I'll quadruple your income. I am going to give you ⅔ of my law firm for free. This is the good news. The bad news is that you won't see much of me. I am going to Belize.'

"They could use my name, my reputation – I would help, but I wouldn't be in the office. I go up to court every summer in suit and tie. So I bought a laptop, a copy machine, a fax machine, and a new mask, fins and snorkel, and I moved to Belize and have been here ever since. I spend about three hours a day on the Internet communicating with my office, my clients, other lawyers and judges. I spend about half an hour on the telephone doing the same thing.

"And I spend June, July and August in the US wearing a suit and tie and going into court. Some of the three hours are research. The reason I am able to be here is because of technology. The rest of my time I spend fishing, growing bananas, diving, talking to people who want to write books, and traveling. This is really my base here to travel around Latin America.

Maybe when in four years my daughter is out of high school, we may move to Vietnam, buy a place, and use it as a base to travel through Asia."

Pat is on the Board of Directors of the Chamber of Commerce in San Pedro. He is fighting to preserve as much of this environment as he can.

"I am active in a lot of civic organizations here. I will take my Belizean citizenship civics test soon. Then I will be a Belizean too. Technology enables my daughter to do Internet high school. That enables us to be here."

Pat also wrote the free "Green Guide to Ambergris Caye" because he couldn't find a ferry, water taxi, airplane schedule, or a map. It's out of print now, but can still be found on the wall in the CocoNet Internet café.

"I recommend a business that does not require physical presence. Why wait until you retire? The Internet world and cell phones and faxes make it possible for a huge percentage of the population to earn a living anywhere. First thing I recommend, quit waiting until you retire."

Pat has also spotted several business opportunities.

"Obviously for expatriates, they are not going to be very comfortable surviving in the Belizean economy because they have been spoiled. Figure out how to survive, make your living in the international economy. Find out how to do that here.

"There is tourism; you run dive shops, fishing charters, and hotels, restaurants, real estate or you telecommute like I do, unless you are independently wealthy or retired. There is money to be made in real estate, as a real estate agent."

When Pat was young and single he was happy to come to Belize and wash dishes and dig ditches. He still is.

"That is getting into the local economy. If you are spoiled by first world economies, you have to be young and not have family responsibilities to thrive in a third world economy. But it sure is fun."

San Pedro is the most expensive place in the country.

"If I were looking to survive in the Belizean economy I'd go to some place like San Ignacio or even Corozal, if I were young and single again. I have obligations. If I were 20 years old I'd go to Dangriga or Cayo. The opportunities are where the tourists aren't. That is where your opportunities are, like the Old West in the US. You can buy a piece of land for $500 [USD] and sell it in five years for $10,000. You can't do that in San Pedro anymore. If you want make real investment in San Pedro you have to have a lot of money now.

"There are lots of opportunities in southern Belize, and along the Mexican border in northern Belize. The economy is wide open to development. Wages are low, and so are land prices. Wages are high in San Pedro, and so are land prices.

"I love the beach. If I had $5,000 [USD] I would buy beach land between Dangriga and Placencia, or river land. If I had $3,000 I'd buy land in Monkey River Town, south of Placencia. In 20 years both sides will be Hyatts, Mariotts. Now you can buy a lot there for $1000 on the water or $500 off the water, versus 5,000 or 10,000 for riverfront property in Southern Belize.

"Who wants to be a real estate investor?

"I came down here originally to buy a 42-foot catamaran to live on. It had two bathrooms and four bedrooms. It was retired from the TMM fleet and cost about a quarter of a million dollars US. I was going to close the deal on Wednesday. On Saturday, Hurricane Keith came in and wiped out the whole TMM fleet. The one I was buying is now part of the reef at Hocol Chan.

"I couldn't afford a new one, and all the old ones were gone, so I bought a house instead. My wife and daughter would have been miserable living on a boat."

Pat Stiley made his dream come true and so you can, too. Just visit Belize. (You can contact the author at helga.peham@chello.at for your comments.)

In Memoriam Janet

Along the Lagoon

Overview

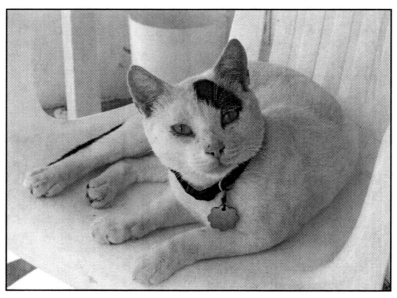

A Cat's Pleasure

Abstract

A Place for You!